Born in San Antonio, Texas, Kitty Ferguson first became interested in mathematics, physics and cosmology when her father went to imaginative lengths to explain these subjects to his wife and children at the family dinner table. Nevertheless, at age 19 she decided to concentrate on studying music and moved to New York where she received bachelor's and master's degrees from the Juilliard School. For many years she was a successful professional musician, conducting and performing oratorio, early music and chamber music.

In 1986, Kitty moved to England, where her husband was a visiting fellow at Cambridge University. During this and many subsequent periods living in Cambridge, Kitty renewed her lifelong interest in physics and cosmology by auditing graduate lectures and seminars at the Department of Applied Mathematics and Theoretical Physics, where she got to know some of the legendary figures in those fields, among them Stephen Hawking. In 1987, she decided to retire from music and devote herself full time to writing about science. Her first book, *Black Holes in Spacetime* (Franklin Watts, 1991), resulted from an award-winning project she helped her eight-year-old daughter create for a school science fair.

In 1989, Kitty approached Stephen Hawking with the idea of writing a book about him and his theories. He had never previously cooperated with authors of other books about himself, but agreed to do so with Kitty, going over some of the chapters and providing her with personal photographs and previously unpublished material. *Stephen Hawking: Quest for a Theory of Everything* (Franklin Watts, 1991; Bantam Books, 1992) was a Sunday Times bestseller and is now available in 15 languages. With *Prisons of Light – Black Holes*, Kitty returns to the subject that first drew her into writing about science.

What is a black hole? How does it 'work'? Could we survive a visit to one . . . perhaps even venture inside? What would we find? Have we yet discovered any real black holes? And what do black holes teach us about the mysteries of our Universe?

These are just a few of the tantalizing questions examined in this tour-de-force, jargon-free review of one of the most fascinating topics in modern science. In search of the answers, we trace a star from its birth to its death throes, take a fabulous hypothetical journey to the border of a black hole and beyond, spend time with some of the world's leading theoretical physicists and observational astronomers scanning the cosmos for evidence of real black holes, and take a whimsical look at some of the wild ideas black holes have inspired.

Prisons of Light – Black Holes is comprehensive and detailed. Yet Kitty Ferguson's lightness of touch, her down-to-earth analogies, and her desire to bring the excitement of science to a wide audience set this book apart from all others on black holes and make it a wonderfully stimulating and entertaining read.

Also of interest in popular science

The Thread of Life
SUSAN ALDRIDGE

Remarkable Discoveries
FRANK ASHALL

Evolving the Mind
GRAHAM CAIRNS-SMITH

Extraterrestrial Intelligence
JEAN HEIDMANN

Hard to Swallow: a brief history of food
RICHARD LACEY

An Inventor in the Garden of Eden
ERIC LAITHWAITE

The Clock of Ages
JOHN MEDINA

Beyond Science
JOHN POLKINGHORNE

Improving Nature? The science and ethics of genetic engineering
MICHAEL REISS & ROGER STRAUGHAN

KITTY FERGUSON

Prisons of Light – Black Holes

Published by the Press Syndicate of the University of Cambridge
The Pitt Building, Trumpington Street, Cambridge CB2 1RP
40 West 20th Street, New York, NY 10011-4211, USA
10 Stamford Road, Oakleigh, Melbourne 3166, Australia

First published 1996

Printed in Great Britain by Biddles Ltd, Guildford & King's Lynn

A catalogue record for this book is available from the British Library

Library of Congress cataloguing in publication data

Ferguson, Kitty.
Prisons of light: black holes/by Kitty Ferguson.
p. cm.
Includes index.
ISBN 0 521 49518 0
1. Black holes (Astronomy) I. Title.
QB843.B55F54 1996
523.8′875–dc20 96–11729 CIP

ISBN 0 521 49518 0 hardback

To Colin and Duff

Contents

Acknowledgements

The author wishes to thank the following for their encouragement and help, for reading over and checking portions of this book, discussing their research with her, providing her with relevant articles, papers, and drawings, and keeping her up-to-the-minute on both the observational and the theoretical sides of black hole research:

Lola Chaisson, Laird Close, Sidney Coleman, Joseph Dolan, Andrew Dunn, Emilio E. Falco, Caitlin Ferguson, Holland Ford, Neil Gehrels, Kim Griest, Carole Haswell, Stephen Hawking, Tod Lauer, Jeffrey McClintock, Yichuan C. Pei, Kenneth Pounds, Fred Seward, Joseph Taylor, Kip Thorne, William Unruh, John A. Wheeler, Clifford Will.

Prologue

For many hundreds of thousands of years, we who live on the earth have watched the night sky, awed by its beauty and mystery, longing to understand what we see . . . and to see more. For most of that time we had only our eyes, our imaginations, and our mathematics to help us, until in 1609 the first telescopes appeared and opened up vast new possibilities. Over the next three hundred years we improved our telescopes dramatically and formed the impression, coming entirely from visible light, of an elegant universe, predictable and patterned, not changing discernably over millions or even billions of years.

There was much we were missing.

The earth's atmosphere is a barrier. By the 1930s we were able to use radio receivers to discover things that aren't visible with any optical telescope, but we were still like crabs scuttling on the ocean floor, wondering what we would find if we stuck our heads above the surface of the water. So we proceeded to put X-ray detectors first on rockets and, in the 1970s, in orbit beyond our atmosphere and found much more in the heavens than anyone could see on a clear night with the most powerful earthbound telescope.

With radio and X-ray telescopes and later with infrared and gamma-ray telescopes, we've discovered that the universe is not, after all, the serene universe we thought we knew early in the twentieth century. It's much more violent and complex, less easy to predict, with stars ripping hot gas from their neighbours, beams of radiation sweeping around from swiftly rotating pulsars, cataclysmic explosions in the cores of galaxies, quasars changing in brightness over very short periods of time, and jets of gas spewing out over tremendous reaches of space from the nuclei of galaxies and quasars.

At the close of the twentieth century we see far more than our predecessors did a mere seventy years ago, but even with powerful telescopes probing the universe in every part of the spectrum from the earth's surface, from orbit, and from space probes, we still don't see everything we know is out there.

No one has seen a black hole, nor, it seems, will anyone ever see one. Until the late 1980s no one could even claim to have found conclusive indirect evidence of a black hole. However, with the eyes not of telescopes but of physics theory and mathematics we have been scrutinizing these amazing phenomena for a long time.

The idea that there might be 'dark stars' with gravitational attraction so powerful that all their light is pulled back in is not a product of twentieth-century science. It came from the British natural philosopher John Michell in 1783, about a hundred years after Isaac Newton introduced his theories of gravity. That is not to suggest that scientists have been thinking about the notion for two hundred years. On the contrary, the idea languished in almost total obscurity from about twenty-five years after its introduction until the second decade of the twentieth century, when Albert Einstein gave us a deeper understanding of gravity and theorist Karl Schwarzschild became curious about how a star's gravity, in Einstein's new way of thinking about it, might affect its light.

By the mid-1980s the theory of black holes was highly respected and meticulously developed physics theory. It was taught in most graduate schools. Confidence ran high that black holes exist. There were several good black hole 'candidates'. But unequivocal evidence of a black hole continued to elude astronomers. Now, a few short years later, we do have convincing observational evidence of the presence of several black holes and good reason to conclude there are many more. The evidence is indirect evidence – circumstantial evidence it would be called in a court of law. No telescope has shown us a picture of a black hole. Discovering a promising candidate and demonstrating that this candidate actually *is* a black hole requires an ingenious collaboration of theory and observational astronomy. Nevertheless, in the case for and against black holes, the jury has returned a verdict. These unlikely, fantastic creatures do exist. We have indeed discovered several.

Does the black hole saga end here? Shall we tuck it away in the history books, enshrine it in permanent science museum exhibits, turn our attention to other matters? The theorists, the observational astronomers, and the popular media are satisfied . . . and that's it? No. We've only begun to find out what black holes are like and what roles they play in the real universe, and we do not yet understand what happens at their centres. Hidden within black holes are clues to what physicist John A. Wheeler calls the 'deep, happy mysteries' of the universe. The attempt to unlock these mysteries is the great adventure of modern science. Black holes don't give up their secrets easily. Finding real black holes hasn't allowed us to probe inside them. We may never be able to do that.

What are these inscrutable, invisible objects? Where do they come from? What is the source of their incredible power? What makes us so certain they are there, if we can't ever actually see them?

1

A cosmic case of burnout

The Great Nebula in Orion, a region of star formation, is visible to the naked eye as the 'star' in the 'sword' of the constellation Orion. (National Optical Astronomy Observatories.)

It is therefore possible that the greatest luminous bodies in the universe are on this very account invisible.

Pierre-Simon Laplace, 1795

Gazing at the night sky may indeed give us the impression that stars are eternal and unchanging. They are not. A star is born and shines for millions, perhaps even billions of years, and then it stops shining. Less massive stars like our own sun burn longer than more massive stars. The lifetime of the sun is likely to be ten to twelve billion years in all. But no star shines forever.

A black hole is one of the ways a star may spend its old age, one possible 'final state' of a star. Though not every star ends up as a black hole, and not every black hole began as a star; nevertheless, if you want to understand black holes, you need to know something about stars.

Figure 1.1 is an overview of the life of a star like the sun, a story that begins in the cloud of gas at the left centre of the picture and ends with the thin line trailing off on the right. This particular star doesn't become a black hole. The artist has drawn the thin line to indicate that it spends its old age as a 'white dwarf'. If it were a black hole, we would see nothing at all trailing off on the right. It may reassure the faint-hearted to know that according to our best knowledge about stars and black holes the sun will never become a black hole. We'll learn why in the pages that follow. However, the mass of the sun is important in the study of black holes because it has become the standard of measurement when we talk about how massive a star or a black hole is. If you read about a black hole of 'one hundred solar masses', that means its mass is one hundred times the mass of the sun. For those whose memories are vague when it comes to the meaning of 'mass', there will be a briefing in a moment.

Enormous clouds of gases out in space, like the one represented in the life-cycle drawing, are the nurseries of stars. Those of us who prefer to take nothing as a given may wonder why we should

The life-cycle of a star like the sun – from its birth in a cloud of gas (left centre) to its old age as a white dwarf (represented by the thin line trailing off on the lower right).

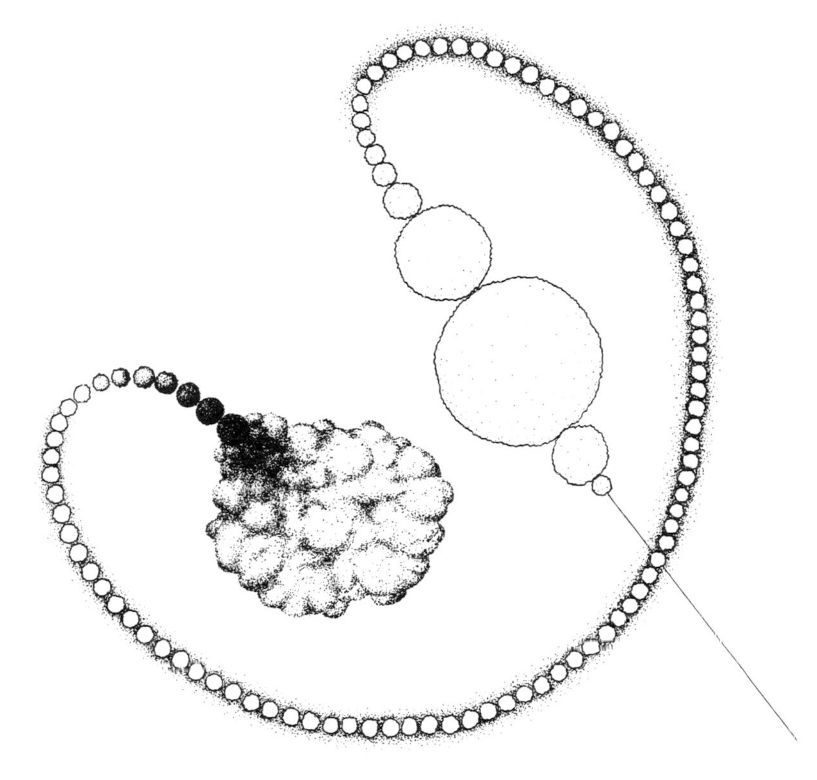

Figure 1.1.

begin the story there rather than ask where the cloud of gas came from – a question that currently occupies some of the foremost minds in physics and astrophysics. There have been theories and discoveries that show hope of unravelling the mystery of why gas or anything else has bunched together in one area rather than remaining evenly dispersed throughout the universe. Most promising, in the spring of 1992, George Smoot of the University of California at Berkeley and colleagues discovered wrinkles in the cosmic background radiation originating in the early universe, wrinkles that were an early indication of such unevenness. The subject is, regrettably, outside the scope of this book. We must begin our story in an era when we know matter *has* accumulated because we have observed it.

A good example of an accumulation of gas is in the magnificently photogenic Great Nebula in Orion (Chapter 1 frontispiece). The gas there is mostly hydrogen. Hydrogen gas, like all other ordinary matter in the universe, consists of atoms. Atoms in turn are composed of elementary particles and a comparatively enormous amount of empty space. The familiar schoolbook drawing of an atom (Figure 1.2*a*) shows a nucleus in the centre made up of par-

(*a*) The Rutherford model of a helium atom.

In this model, the electrons orbit the nucleus the way planets orbit the sun, in well-defined paths.

Notice that there is a lot of empty space in an atom - much more actually than this drawing suggests, because even at this magnification the electrons and the nucleus are so small that you really couldn't see them at all.

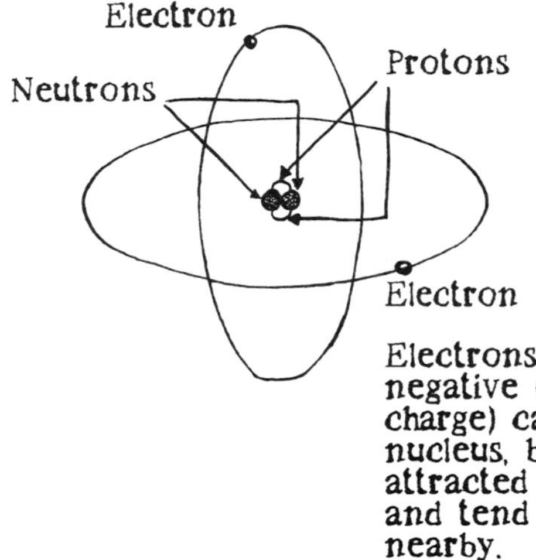

Protons (with positive electrical charge) and neutrons (with no electrical charge) cluster together at the centre of the atom, in the nucleus.

Electrons (with negative electrical charge) can't be in the nucleus, but they are attracted to protons and tend to stay nearby.

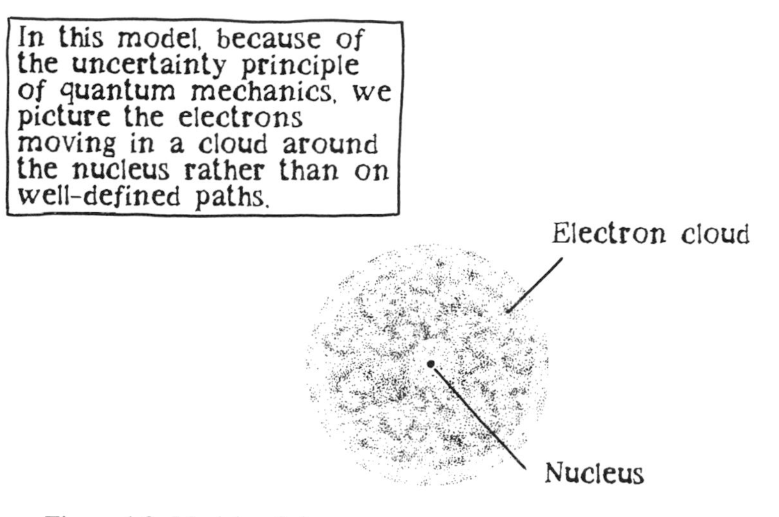

Figure 1.2. Models of the atom.

ticles known as protons and neutrons, with other particles called electrons orbiting the nucleus. That's the way Ernest Rutherford drew the atom in the early part of the twentieth century. In order to understand some parts of this book, it will be necessary to relinquish that familiar picture and replace it with something less easy to draw or comprehend, more like the drawing in Figure 1.2*b*, but neither drawing is adequate when it comes to scale. We'll be closer to the correct scale if we imagine the nucleus being about the size of a city block and the electrons orbiting or swarming in a cloud in an area about the size of the earth. Also, neither drawing even begins to indicate how much empty space is in an atom, and this empty space plays a key role in the fate of a star.

In an area such as the Orion Nebula, the atoms and particles have come near enough to one another in a gas to begin pulling on one another and drawing closer together still. We call this pull 'gravity', or gravitational attraction, something we take for granted as the force that keeps us from falling off the earth. We think less often of it working among the tiniest particles and among all the

An example of **nuclear fusion:** A deuterium ('heavy hydrogen') nucleus, which contains one neutron and one proton, and a hydrogen nucleus, which consists of a single proton, get fused together to create a helium-3 nucleus, which contains two protons and one neutron. This is one of the fusion reactions that power the sun and other stars.

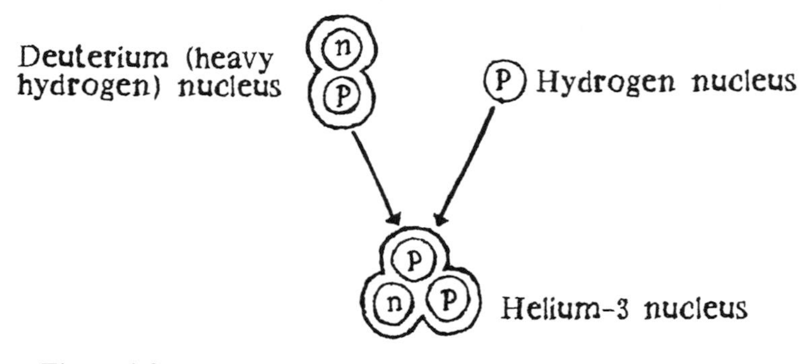

Figure 1.3.

enormous objects out in space. Much of this book will be about the way gravity operates. We begin simply by noting that, other things remaining equal, the nearer objects approach one another the stronger the gravitational attraction between or among them becomes. In the hydrogen gas we've been talking about, the atoms respond to the pull of one another's gravity and draw closer yet, moving faster because of the stronger gravitational attraction. Collisions occur more and more often and at higher speeds. Everything heats up. When things have become sufficiently hot, the atoms in the gas stop bouncing off one another. Instead, the hydrogen nuclei begin to merge into one another and change into helium. This is the process called nuclear fusion (Figure 1.3). Because of the overcrowded conditions among the atoms in certain parts of the cloud of hydrogen gas, there will be a lot of it going on. It's like a controlled hydrogen bomb explosion, and the heat is indeed intense. We're no longer looking merely at a cloud of gas. Some of the gas has pulled itself together to form a star. The release of heat, as hydrogen nuclei continue to fuse, makes the star shine.

Stars spend most of their adult years converting hydrogen to

Gravity vs pressure from the heat released in nuclear reactions

A competition between two closely matched teams keeps a star shining for millions or even billions of years.

Pressure from the heat released in nuclear reactions (dark arrows) supports the star against the squeeze of its gravity (light arrows).

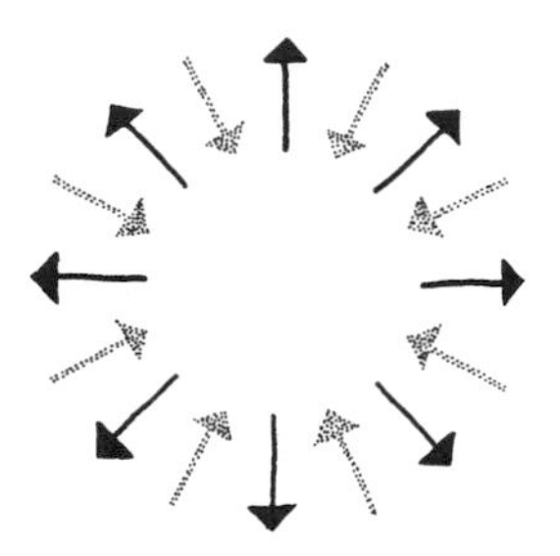

Figure 1.4.

helium, and that's what keeps them alive. For all those millenia there is a balance – a tug-of-war, if you will – between two closely matched opposing teams. On the one hand there is gravity, the force that continues to draw the atoms and particles closer together. Left unopposed, gravity would cause the star to collapse in on itself. We who depend on a star for our continued existence are fortunate that gravity does not go unopposed. Paradoxically, gravity brings about nuclear fusion and thus assures that there IS an opposing team. The heat released in the nuclear reactions (as hydrogen becomes helium) creates enough pressure in the gas to balance the gravitational attraction, and the star doesn't collapse. Think of the way the pressure in a balloon holds the walls of the balloon apart. The rubber walls pull toward each other, trying to come together – which they will do quite readily if we let the gas out of the balloon. The pressure in a star doesn't allow it to collapse – which it would do quite readily if the nuclear reactions stopped occurring and gravity were allowed to have its way. The heat released in nuclear reactions also makes up for the heat lost as the star radiates light into space.

Picture this tug-of-war, then (Figure 1.4). Pressure from the heat released in nuclear reactions is one team, gravity the other. Occasionally gravity appears to be winning. When the accumulation of helium ash snuffs out the central furnace of the star, the core begins to collapse, but the collapse itself soon produces higher temperatures and the star reignites, now fusing helium into carbon atoms and swelling into a red giant. In the life-cycle drawing in Figure 1.1 we see that happen. Were the star in the drawing our sun, in the red giant stage it might engulf the earth. In some stars the snuffing out and reignition occur several times, with stars fusing atoms into heavier and heavier elements.

This competition eventually ends with gravity the victor. When most of the hydrogen atoms have fused into helium atoms and some of those into carbon atoms and, in more massive stars, into heavier elements, the star finally exhausts its usable fuel. Nuclear reactions happen less and less often as the fuel runs out, but the star continues to radiate light into space even when it is no longer able to make up for the heat lost in this radiation. There is less and less heat and pressure to counteract gravitational attraction. After patiently holding its own for millions or billions of years, gravity wins the tug-of-war. The cooling star begins to shrink and collapse in on itself.

A very massive star during these last stages may blaze into a giant Type II Supernova, shining for a few weeks as brightly as ten million suns. While its centre collapses and crunches down to the density of an enormous atomic nucleus, its outside rips apart and becomes a glowing, expanding shell of gas. In the explosion the star throws off some of its matter, sometimes a very large portion of it, thus reducing its mass before continuing to collapse. How great a mass it ends up with after the explosion will be instrumental in deciding its fate.

It would perhaps be prudent to digress for a moment and clear up any confusion about the meaning of 'mass'. We should not confuse mass with size. A small object can be 'massive'. Within the earth's gravity, we translate that to mean that a small object can be heavy, which is obviously true. A popular children's book about black holes makes the point with an illustration showing a mother and child watching a see-saw. At one end of the see-saw is a stack of elephants, each elephant standing on the back of the one below, stretching up out of sight at the top of the page. At the other end of the see-saw is a tiny pebble. We can tell from the way the see-saw is tilted that the pebble is outweighing the tower of elephants. Elementary school children learn that mass is a measure of how much matter (or how many particles of matter) is in an object, regardless of how loosely or densely that matter is packed. In the see-saw drawing, we know the pebble must consist of many more particles of matter than the tower of elephants does. The particles of matter in the pebble are much more densely packed. A more sophisticated definition of mass is that mass has to do with how much an object resists any attempt to change its speed or direction, but for our immediate purposes the important thing to understand is that as a star shrinks and becomes smaller it does not necessarily become less massive. An aged, drastically shrunken star that has not thrown off mass in a supernova may be as massive as it was in the prime of life.

Three possible fates now await a collapsing star.

1. It may settle down and spend its old age as a 'white dwarf' like the star in the life-cycle diagram, with a radius of a few thousand miles, not too much smaller than the earth but with a density of hundreds of tons per cubic inch. We observe many white dwarfs in our own part of the Milky Way galaxy. There is one orbiting Sirius (the Dog Star), the brightest star in our night sky (Figure 1.5). As we've said, the sun will probably end up as a white dwarf.
2. It may not settle as a white dwarf but instead crunch down until its circumference is approximately a mere 100 kilometres, ending up as a 'neutron star' with an incredible density of millions of tons per cubic inch. 'Pulsars', which we shall encounter later in this book, are neutron stars.
3. It may continue to collapse and form a 'black hole'.

Which will it be? What causes some stars to retire as white dwarfs, while others collapse to neutron stars and still others are doomed to crunch all the way down to black holes?

After all the fuel for nuclear reactions has been used up, gravity meets a fresh opponent – the 'exclusion principle'. All particles of ordinary matter – the particles that make up atoms, which in turn make up such things as this book, you and I, stars, gases – obey this principle. The exclusion principle insures that particles keep their distance and there will be empty space in atoms. Without it we would have no stars, or people, or any other of the familiar

Figure 1.5. Sirius B, a white dwarf seen here as a small dot of light, orbits Sirius A, the Dog Star, the brightest star in our night sky. The spikes in the picture aren't real: they are caused by the support struts of the telescope. (Lick Observatory Photo/Image.)

objects in our universe. We would have something more like a dense soup.

The exclusion principle requires that no more than two electrons can occupy the same region of space at the same time. Electrons – moving in clouds around the nuclei of atoms – are paired together in cells (or 'orbitals'). Electrons protest their confinement in these cells by moving erratically, shaking, flying around, kicking forcefully against adjacent electrons. This motion is called 'degenerate motion', and the pressure it produces is 'electron degeneracy pressure'. This pressure keeps electrons from being pulled into the nucleus of the atom.

As a star collapses, the clouds of electrons around the nuclei of the atoms in the star get squashed until the electrons are confined in cells many times smaller than they usually could move around in. In this situation an electron begins to behave in part like a wave. That may seem unlikely, but I will ask you to accept it for now and look forward to an explanation later. It stands to reason that the length of the wave cannot be larger than the cell the electron is in. Shorter wavelengths mean higher energy. Higher energy implies more rapid motion. It follows that the denser the matter in the star becomes, the smaller the cells will be; the smaller the cells, the shorter the electrons' wavelengths must be; the shorter the wavelengths, the higher the electrons' energy; the higher the energy, the faster the electrons' motion; the faster this motion, the larger the electron degeneracy pressure it produces. It is this pressure that will continue to support the star against gravity.

This scheme sounds as though it would continue to work in a highly satisfactory manner. As the pull of gravity increased, so would the pressure opposing it. There is, however, a hitch. Our universe has a speed limit which for all practical purposes seems to be unbreakable. That speed limit is the speed of light, approximately 300,000 kilometres (or 186,000 miles) per second. Degenerate electrons can't move faster than the speed of light. But even short of that, when matter is so dense that degenerate electrons move at *near* the speed of light, matter has serious difficulty supporting itself against the squeeze of gravity. Can it succeed?

In the late 1920s Subrahmanyan Chandrasekhar, a young Indian physicist then at the University of Cambridge, calculated that if a star's mass is less than 1.4 times the mass of our sun, gravity will not be able to overpower this exclusion principle repulsion among the electrons. The star shrinks and becomes a white dwarf, but, because of the exclusion principle, it shrinks no further. That star will not become a black hole. We now call this mass of 1.4 solar masses the 'Chandrasekhar limit'. Only in a star whose mass is *more* than the Chandrasekhar limit will gravity overcome the exclusion principle among the electrons and be the victor in this second competition (Figure 1.6).

Let's suppose now that we are dealing with a slightly more massive star, more than 1.4 solar masses. Is every star with a mass over the Chandrasekhar limit destined to be a black hole?

The Russian scientist Lev Davidovich Landau, who arrived at a limit similar to Chandrasekhar's, also called attention to another possible final state for a star. When gravity has overpowered the exclusion principle repulsion among the electrons, as the star

Gravity vs the exclusion principle

After all the fuel for nuclear reactions is used up, gravity meets a fresh opponent - the 'exclusion principle'.

Exclusion principle repulsion (dark arrows) among the electrons (in a white dwarf) or among the neutrons (in a neutron star) supports the star against the squeeze of its gravity (light arrows).

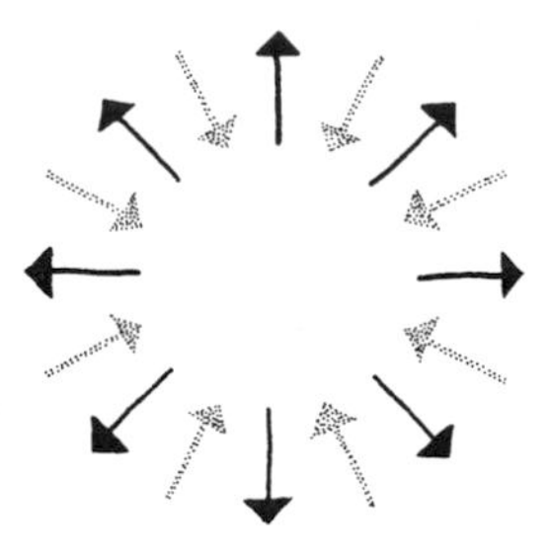

Figure 1.6.

squashes down, electrons are squeezed into the atomic nuclei and combine with protons in the nuclei to form additional neutrons. After a while the core of the star is almost entirely made of neutrons. Neutrons must obey the exclusion principle as surely as electrons must. The resistance to squeezing, partly due to degeneracy pressure and partly due to another force called the strong nuclear force, is stronger than it was previously among the electrons, presenting gravity with an even more formidable opponent. There is a complicating factor here, and no one is quite sure how important it might turn out to be. It's possible that the electrons in the crunch won't combine with protons but will instead turn into other particles which *can't* combine with protons, leaving plenty of protons hanging around among the neutrons. A mixture of protons and neutrons is easier to compress and the star is more likely to form a black hole. Nevertheless, the current consensus is that the maximum allowed mass for a 'neutron star' lies between 1.5 and 3 solar masses and is most likely to be about 3 solar masses. A neutron star is smaller and denser than a white dwarf.

However, suppose we have a star a little more massive than the maximum allowed mass for neutron stars. Is *that* star destined to become a black hole? Not necessarily. A star may lose a considerable amount of mass in a late-in-life explosion. However, if it doesn't lose enough to bring it below the maximum allowed mass for neutron stars, about 3 solar masses, gravity *will* overpower both the exclusion principle repulsion among the neutrons and the strong nuclear force, and the star will continue to collapse. When it has reached a size not much smaller than it would have been had it remained a neutron star, the star will form a black hole.

Common sense suggests that there is something suspicious about this description. Why should it be the *more* massive stars that end up smallest, as black holes, and *less* massive stars that end up larger, as white dwarfs or neutron stars? Shouldn't it be the other way around? That question gives us a good excuse for a closer look at gravity.

Physicist John A. Wheeler of Princeton and the University of Texas asks us to think of gravity as a universal democratic system, with every particle in the universe casting a vote that can affect every other particle in the universe. Though no particle of matter has much gravitational influence by itself, when particles join forces and vote as a bloc (the earth, for instance, or a star), that bloc can wield enormous influence. The combined votes of all the particles in this great hulk of a planet under our feet obviously constitute a significant amount of gravitational clout.

We've said that mass is a measure of how much matter is in an object, how many matter particles have banded together to form this particular voting bloc. The greater the number of votes in the voting bloc, the greater the mass, and the greater the influence of that voting bloc. Getting back to the stars whose life stories we have been tracing – it should be clear that saying one star is more massive than another means that there are more particles of matter in it. The more particles of matter, the more gravitational attraction. Hence, the more massive the star, the more powerful the gravity team is in the tug-of-war, the more powerful the potential 'squeeze', and the more likely it is that the squeeze will be able to overpower the exclusion principle.

Gravity was the force that gave birth to the star in a cloud of gas. It was the force that kept the star in balance for all those years, not allowing it to fly apart but also contributing to the processes that prevented its collapse. Now, after millions or even billions of years have passed, gravity is the force that claims the star as its victim, squeezing all its enormous mass to something the size of the sun . . . then to the size of the earth . . . to the size of the moon . . . to the size of London . . . St. James Park . . . the lake in the park . . . a duck on the lake . . . a tennis ball . . . a marble . . . the head of a pin . . . the point of a pin . . . a microbe . . . For a star ending up more massive than about 3 solar masses, we know of no power capable of halting this catastrophic collapse. The star continues to crunch down even after it has become a black hole. It's best to think of a black hole *not* as a star, but as what happens to spacetime around a star that goes on collapsing to near infinite density.

2

Matters of gravity: Newton and Einstein

Left: Isaac Newton. Born 25 December 1642, Woolsthorpe, Lincolnshire, England. Died 20 March 1726, Kensington, London. (American Institute of Physics, Emilio Segre Visual Archives, W.F. Meggers Collection.)
Right: Albert Einstein. Born 14 March 1879, Ulm, Germany. Died 19 April 1955, Princeton, New Jersey. This photograph was taken in 1931. (The Archives, California Institute of Technology.)

Great thinkers of the past, you identified and solved many a mystery
of motion where others could see no mystery . . .
Thanks to your hard-won insights, the whole great story of gravity
Now comes to us in a single simple sentence:
Spacetime grips mass, telling it how to move;
And mass grips spacetime, telling it how to curve.

John Archibald Wheeler

We've just seen gravity crush a dying star to a degree beyond the limits of human imagination. As we continue, we'll find that it's often difficult to accept the peculiar and counterintuitive things that occur where gravity is enormously strong. You might be well advised to adopt, early on, the attitude of the White Queen in *Alice Through the Looking-Glass*, for whom it was routine to believe 'as many as six impossible things before breakfast'. Nevertheless, in this chapter we're going to see that the laws by which gravity operates – as we humans are able to understand those laws at present – the laws that underlie much of the mind-boggling extravaganza of this cosmos, are strikingly simple and ingenious.

We'll begin our approach to them with a bit of science fiction (Figure 2.1).

First, take note of what the pull of gravity feels like on the surface of the earth as we know it today. Then, pretend that you board a spacecraft and leave on a short holiday in space. During your absence something bizarre happens to the earth. It gets squeezed to half its former size. It still has the same mass it had before, but that mass is pressed together more densely. Approaching the earth on your return journey, after the squeezing, you arrive where the surface of the earth used to be and find yourself still in space with some distance to travel before reaching the new surface which is considerably 'further in'. Let us suppose you stop to hover nostalgically where the old surface used to be – an imaginary surface that no longer exists except in memory. What does the pull of gravity feel like to you there? Surprisingly, it feels exactly as

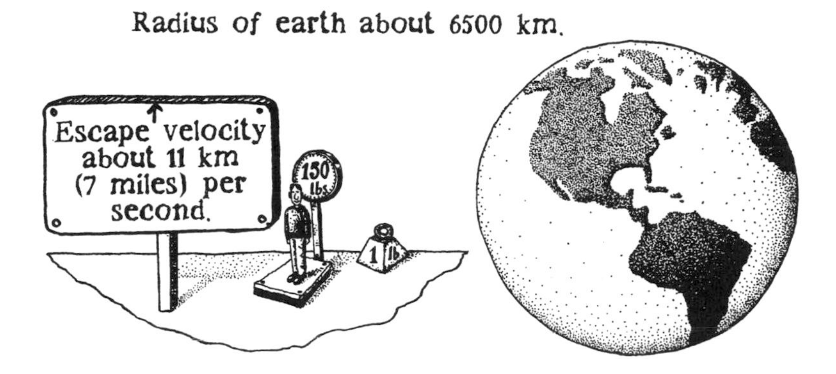

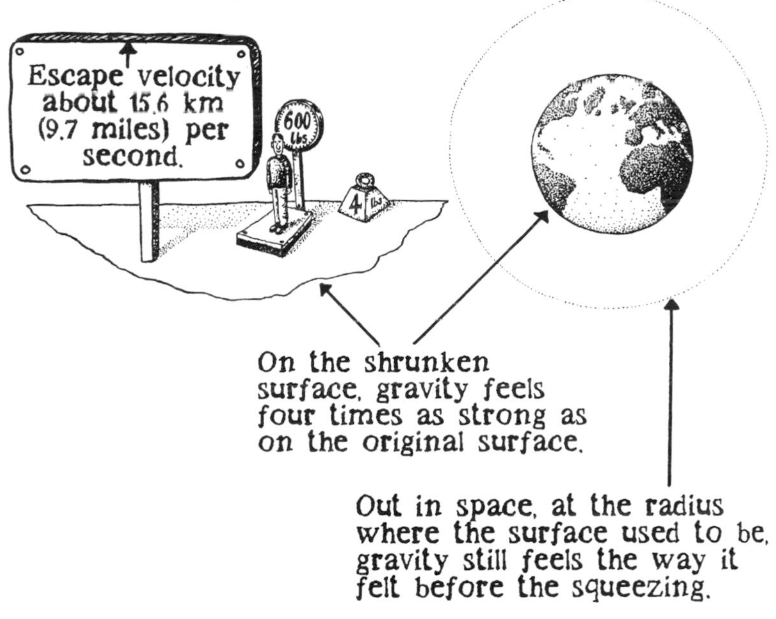

Figure 2.1. The day the earth was squeezed.

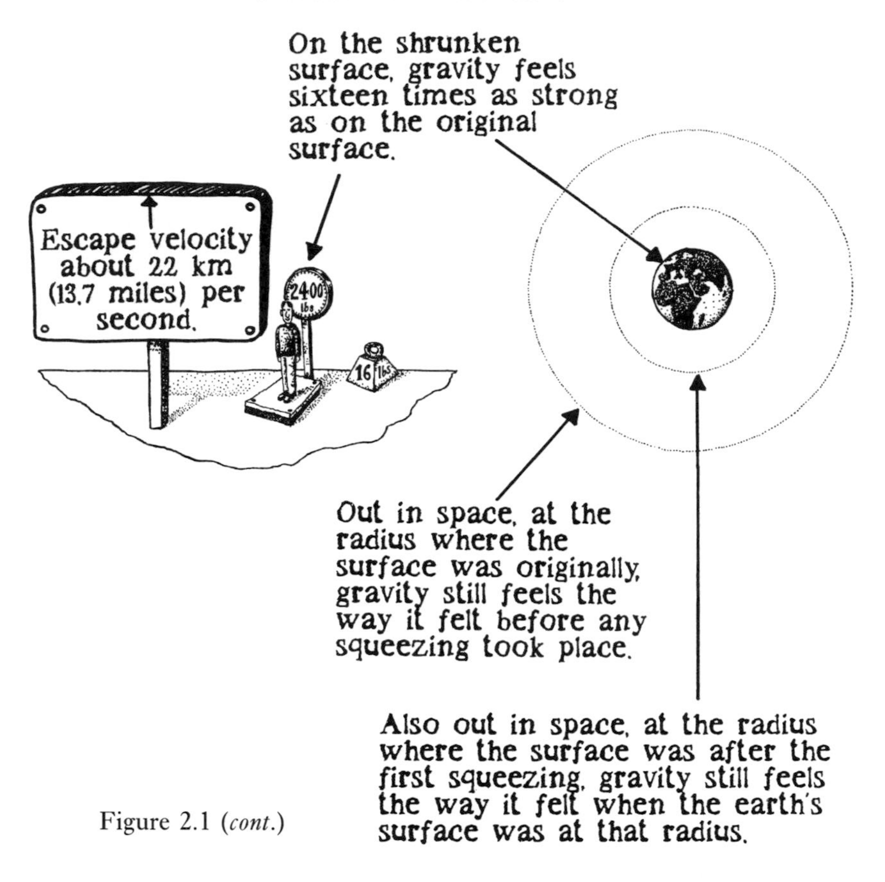

Figure 2.1 (*cont.*)

gravity felt to you on the surface of the earth before the squeezing occurred. You feel as heavy as you did before you went away on your holiday. You notice that the moon, out beyond you, still orbits as it did before. However, when you reach the new, shrunken surface, the gravity on that surface is four times what you recall on the earth's surface before the squeezing. You feel much heavier on the new surface.

Suppose you take another holiday and return to find the earth has once again been squeezed. This time you find it reduced to

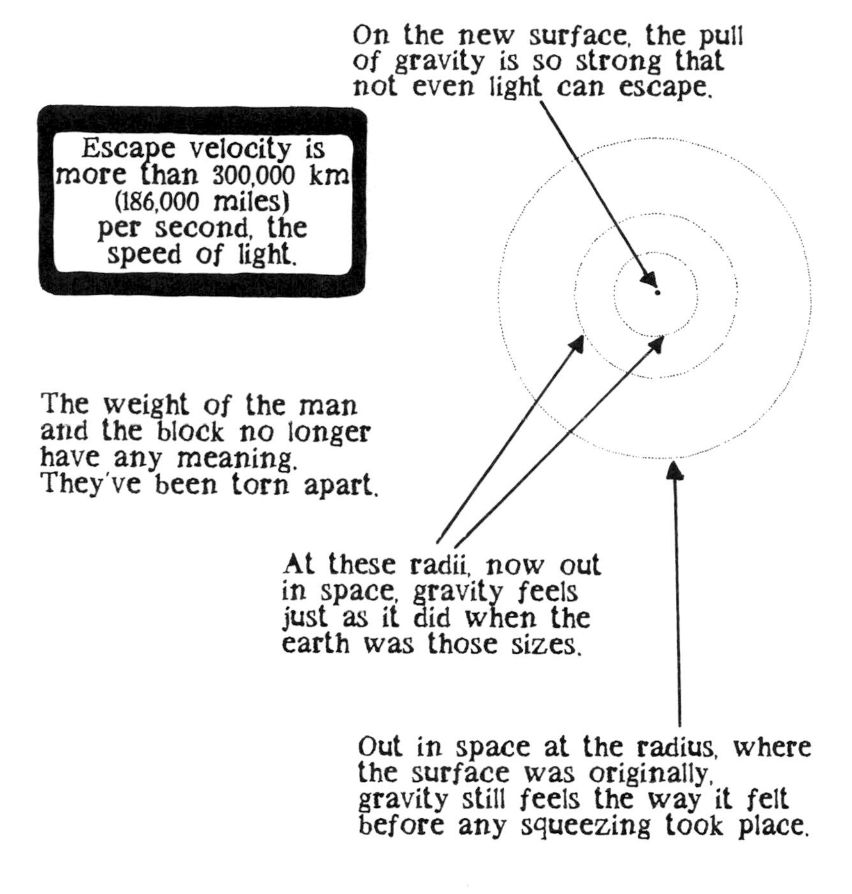

Figure 2.1 (*cont.*)

only a quarter of its original diameter, but still its mass hasn't changed. Everything is just crushed together more densely. Once again you hover for a moment where the surface was before any squeezing at all occurred, and once again you find that gravity there feels just as it did in the good old days. You may begin to suspect that regardless of how much the earth gets squeezed and at all stages of the squeezing, as long as the mass of the earth stays the same, you can have your moment of nostalgia out there in space where the original surface had been, because the pull of gravity out there is not going to change. Your suspicion is correct.

However, you will want to find out what the new surface feels like, so you continue your journey down. It probably won't surprise you that when you reach the place (now out in space) where the surface was after the first squeezing, gravity at *that* point feels just the way it did after the first squeezing. On the newest surface, the pull of gravity is 16 times as great as it was on the original surface.

If the squeezing can happen twice, it might happen again, and it would be prudent to take another holiday in space so as not to be on the earth's surface when the next squeezing occurs. This time, the earth is squeezed until its circumference is that of a pea – all the mass of the earth, billions of tons, crushed into that tiny area! Gravity on the pea-sized earth's surface is so strong that the escape velocity is greater than the speed of light. Escape velocity means the velocity at which an object would have to be fired vertically upward from the surface in order not to be dragged back by the gravitational field of the earth. An object fired vertically upward from the surface at a velocity less than the escape velocity cannot escape. When escape velocity is greater than the speed of light, even light can't escape. The earth is a black hole. But out in space at the distance where the original surface was you can hover to your heart's content, experiencing gravity just as you did when you lived on the earth before any squeezing occurred. The moon still orbits as though nothing untoward had happened to the earth.

What you have felt as the earth was squeezed deserves an explanation, but first let us note parenthetically that, as far as anyone knows, that story is science fiction: it is stars, not planets, that become black holes.

If you find yourself floundering in the next few paragraphs, don't be concerned. Skim them and get on with what comes after. It isn't necessary to understand all this material in order to enjoy reading about black holes. However, if you can follow this discussion and remember some of it as we move ahead, you'll be better equipped to second-guess the experts when we encounter them later in the book. If you're sure you have retained everything you learned about gravity in school physics, you might choose to skip-read much of this chapter.

Isaac Newton, Lucasian Professor of Mathematics at Cambridge in the 1600s, laid the foundations for our understanding of how gravity works in more or less normal circumstances. First, we should bear in mind that bodies (planets and stars, for example) are never to be thought of as 'at rest' in the universe. If you have been thinking they might be, you are in good company. So did Aristotle. However, from Newton we learned that we must not think of objects as sitting still before some force pushes or pulls them, and after the pushing or pulling eventually settling down to sit still again. Instead, a body left undisturbed continues to move in a straight line without changing its speed.

Take the moon, for example: if the moon were alone in space, it would not just sit there like a lemon (nor, for that matter, would a lemon in similar circumstances). It would move in a straight line at a constant speed. Of course, if the moon were truly all alone, there would be no way to decide it was doing this, nothing to which to relate its motion. We now know that it's possible to measure an object's speed and direction *only* in relation to other objects in the universe. We can't measure them in relation to absolute stillness or to anything that resembles absolute north, south, east, west, up or down. But the moon is not alone. The pull of a force we have named 'gravity' influences the moon to alter its speed and its direction. Where does this pull come from? Most noticeably from a nearby voting bloc of particles – a massive object called the earth. The moon resists the change and attempts to keep moving in a straight line. How successful that resistance is depends on how many votes are in the *moon*, how massive the moon is. 'Inertia' is the name given to that sluggishness of a body resisting changes in motion. The moon's gravity also affects the earth. The most obvious result is an indirect effect – a slight stretching out of shape, discernible in the ocean tides.

In Newton's description of gravity, each body in the universe is attracted toward every other body. The more massive the bodies are and the closer they are to one another, the stronger the attraction. It follows that any change in the mass of either the earth or the moon would change the strength of the gravitational attraction between them. For example, if the mass of the earth were doubled, the attraction between the earth and the moon would be doubled.

It also follows that any change in the distance between the earth and the moon would change the strength of the gravitational pull between them. If we were to move the moon to twice its present distance from the earth, the attraction of gravity between the earth and the moon would be only one-fourth as strong as it currently is. All this is summed up in the statement: the gravitational force between any two bodies is proportional to the product of their masses and inversely proportional to the square of the distance between them. For our purposes we need only remember that both the *mass* of the objects and the amount of *distance* which separates them are involved when we ask how great the gravitational pull is between them. How large they are is not of interest.

Keeping that much in mind, we can return to the science fiction story about the squeezing of the earth with fresh understanding – as well as some new confusion – and ask a few questions.

First, why, when you return to earth from your first holiday and pause to hover at the place, now out in space, where the earth's surface was before the squeezing, do you feel the same pull of gravity there that you felt before the squeezing? Why does the moon maintain its orbit, oblivious to the dramatic changes in the size of the earth, obviously responding to the same amount of gravitational pull as before?

Let's ignore the moon for a moment and begin with you, a passenger in the hovering spacecraft – not orbiting, mind you, hovering. We've just learned that, according to Newton's theory, any change in the mass of the earth or of you would change the amount of gravitational pull between them. In our story, there was no change in the mass of either the earth or of you. (We dismiss the possibility that you gained or lost weight while on holiday.) We said that the mass of the earth didn't change even when the earth was squeezed to the circumference of a pea and became a black hole, it was only pressed together more tightly. Also according to Newton's theory, any change in the distance between the earth and you in the hovering spacecraft would change the amount of gravitational pull between them. Did the distance change? It would seem it did – and here is where some confusion might arise. The surface of the earth when it becomes a pea is surely much further from you, the passenger in the hovering spacecraft, than the surface of the earth was in the old days when it was right there under your feet.

In the rule-book by which gravity operates, there is a technicality we have so far neglected to mention: when we speak of 'distance' in Newton's laws, what is important to us is the distance between the *centres* of the bodies in question (or, to put that more precisely, between their *centres of gravity*), not the distance between their *surfaces*. When the earth was squeezed, though the surface of the earth certainly moved increasingly further from where we are accustomed to finding it, the *centre* of the earth did not. The mass of the earth pulls on another body with the same strength as if the earth's mass were all concentrated at a single point at the earth's centre.

If we can digest that, we can also see why, in the story, the moon's orbit doesn't change as the earth shrinks. The mass of the earth and the mass of the moon don't change. And the distance between the earth's centre and the moon's centre doesn't change. (Caution: Do not be drawn into thinking that a passenger on the orbiting moon – or anything else that is orbiting – will 'feel' the gravity of the earth as does a passenger in a hovering spacecraft. More on this later.)

Another question. We've said that the mass of the earth pulls on another body with the same strength as if the earth's mass were all concentrated at a point at the earth's centre. Does that mean the centre of our earth, with so much gravitational attraction emanating from it, is *already* a black hole, even before the earth gets squeezed?

The answer is no. To understand why, we must familiarize ourselves with another of Newton's discoveries.

Picture the sphere of the earth, only make it transparent so that nested within it we can see another smaller sphere (Figure 2.2). From Newton we learn that if I travelled through a shaft from the earth's surface down far enough so that the boundary of the smaller imaginary sphere passed through my body, I would no longer be affected by the pull of gravity originating from any mass of the part of the earth outside the smaller sphere. Only the mass inside the smaller sphere would affect me. True, the mass of the earth outside the smaller sphere would be pulling on me in many

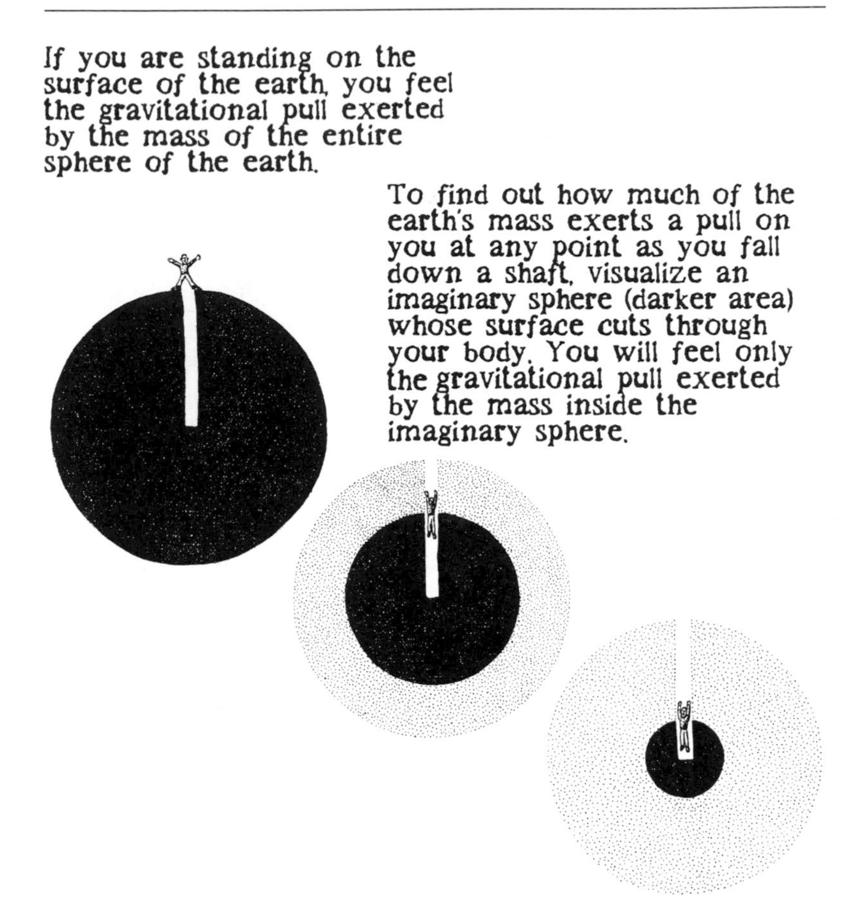

Figure 2.2.

directions and at many strengths (remember, every particle casts its vote), but the net result of that pull on me would be zero. However far we drilled an imaginary shaft toward the centre of the earth, we could imagine the surface of a sphere there (a sphere centred on the centre of the earth), and know that if we positioned ourselves so that the surface of that imaginary sphere passed through our bodies, we would, to all intents and purposes, not feel any gravitational attraction from any of the mass of the earth outside that sphere (Figure 2.2).

Compare that scenario with the end of the squeezed earth story.

If the present earth shrank to the circumference of a pea, without losing any mass, then on the surface of that pea we would feel the gravitational attraction from ALL the mass of the present earth – enough so that the escape velocity would be more than the speed of light. On the other hand, if we were to travel down a shaft to a pea-sized sphere at the centre of the present earth, without any squeezing having occurred, we would feel the gravitational attraction from only the tiny bit of mass inside that pea-sized sphere.

There are two concepts which we should carry away with us from the above story and discussion:

1. Regardless of how small the earth was squeezed – as long as the *mass* of the earth stayed the same – the pull of its gravity at the original radius of the earth (where the surface was before any squeezing occurred) remained the same. This is an important point. Remember it, and you will see that a black hole doesn't sit there sucking in the entire cosmos like a giant vacuum cleaner. Keep a sufficient distance away and a black hole won't affect you any more than any other body with the same amount of mass.
2. In order to have a black hole, we need a great deal of mass not just centred, but *concentrated*, in a small area. That is a rough way of saying what we will be able to state more precisely as we continue.

Newton's theory of gravity is a highly successful theory. We know that it breaks down in some unusual circumstances, such as when gravitational forces become extremely strong or when bodies move at near lightspeed. However, the theory served us well for over two hundred years – during which we didn't really expect we'd need to account for such bizarre circumstances – and it was not improved on until early in the twentieth century, when Albert Einstein began to worry about some problems in the theory. One of these problems was: if it's true that the strength of gravity between two objects is related to the distance between them, then if someone took the sun, for example, and moved it to a greater distance from the earth, would the force of gravity between the earth and sun change immediately?

The speed of light measures the same regardless of where we

are in the universe and regardless of how we're moving – and Einstein recognized that nothing can exceed the speed of light. Light from the sun reaches the earth in approximately eight minutes, not sooner. If the sun blinks out as you read this sentence by daylight, you can go on reading for eight minutes. You will surely finish the chapter. By the same token, if someone moved the sun further from the earth, the earth would not find out about the move or feel any effect of the move for eight minutes. For eight minutes the earth would continue to orbit just as though the sun had not moved. Why? Because gravity also cannot travel faster than the speed of light. The effect of the gravity of one body on another cannot change instantaneously because gravity cannot move faster than about 300,000 kilometres (186,000 miles) per second. Information about how far away the sun is takes *time* to move across space and reach the earth.

In 1908 the German theoretical physicist Hermann Minkowski first introduced the idea of four-dimensional spacetime. To Einstein also it was obvious that when we study the activity of objects in the universe, speaking only in terms of the three space dimensions is inadequate and misleading. If information cannot travel faster than lightspeed, objects at astronomical distances don't exist for us or for each other without a time factor, and describing the universe in three dimensions is as ineffective as describing a cube in two. We must recognize the time dimension as integral to the picture and speak of four-dimensional 'spacetime'. However, that recognition alone didn't solve the problem. In 1915, Einstein in his general theory of relativity suggested we think of gravity not as a force acting between bodies but in terms of the shape or curvature of four-dimensional spacetime itself. In general relativity, gravity is the geometry of the universe.

In an article titled 'Quantum Gravity', in *Scientific American*, December 1983, Bryce DeWitt of the University of Texas offers us a good introduction to thinking about spacetime curvature. He asks us to imagine that we are trying to draw a grid on the earth, not realizing that the earth's surface is curved rather than flat.

> The result can be seen from an airplane on any clear day over the cultivated regions of the Great Plains. The land is subdivided by

> east–west and north–south roads into square-mile sections. The east–west roads often extend in unbroken lines for many miles, but not the north–south roads. Following a road northward, there are abrupt jogs to the east or west every few miles.

DeWitt explains that in attempting to draw a grid, land surveyors were forced by the curvature of the earth to introduce jogs. 'If the jogs were eliminated, the roads would crowd together, creating sections of less than a square mile.' Applying the same principle with more dimensions:

> One can imagine building a giant scaffold in space out of straight rods of equal length joined at angles of precisely 90 degrees and 180 degrees. If space is flat, the construction of the scaffold would proceed without difficulty. If space is curved, one would eventually have to begin shortening the rods or stretching them to make them fit.

According to Einstein's general theory of relativity, the presence of mass or energy causes spacetime to curve, the greater the mass/energy the greater the curvature. Any entity that has any mass at all curves spacetime near it and contributes to the overall curvature of spacetime. So does the presence of energy. Earlier we spoke of every particle of matter having a vote. Are we extending the franchise? Not exactly. Einstein's special theory of relativity taught us that mass *is* a form of energy, an extremely compact form. An 'equivalence' of mass and energy in Einstein's equation $E = mc^2$ (E is energy; m is mass; c is the speed of light) allows us to think of mass and energy as two forms of the same thing – and the presence of this 'thing', whichever form it takes, causes spacetime to curve.

How then does this curvature affect the way objects move in the universe? Objects going 'straight ahead' find themselves following curved paths. One way to get a grip on the idea is to imagine an automobile with the steering locked so that it can only move 'straight ahead'. If the road is properly banked, the automobile will follow the curves in the road in spite of its steering being locked on 'straight ahead'. A different analogy has a bowling ball lying in the centre of a trampoline, making a depression in the elastic surface. If we try to roll a golf ball in a straight line past the bowling ball, the golf ball will change direction when it encounters the depression caused by the bowling ball – as surely as the automobile with locked steering changed direction when it met the banked turns of the road. The golf ball on the trampoline may even trace out the shape of an ellipse and roll back in our direction. According to Einstein's theory, something like that occurs as the moon attempts to continue on a straight path past the earth. The earth distorts spacetime as the bowling ball distorts the rubber sheet. The moon's orbit is the nearest thing to a straight line in warped spacetime.

The trampoline analogy also helps with another concept. The heavier the bowling ball, the more curvature it causes in the rubber sheet. Similarly, the more massive the massive object, the more curvature it causes in spacetime. The sun causes more curvature than the earth. Einstein was describing the same phenomenon Newton described. To Newton, a massive object sends out a force, and how much that force affects another object is related to how massive the objects are and to their distance from one another. To Einstein, a massive object warps spacetime. How much the warp affects another object is related to how massive the objects are and their distance from one another. Clearly, a second bowling ball rolling across the rubber sheet is not going to respond to the warp caused by the first bowling ball in quite the same way the golf ball did. The second bowling ball will be causing a great deal of warp itself as it moves along. Also, a golf ball rolled across the trampoline far from the bowling ball will not be as much affected by the depression caused by the bowling ball as is a golf ball rolled nearer the bowling ball.

John Wheeler, with his gift for words, has summed all of this up neatly in the phrase: 'Mass grips spacetime, telling it how to curve. Spacetime grips mass, telling it how to move.'

If we use Newton's theories to calculate planetary orbits in the solar system, and then calculate them again using Einstein's theories, we arrive at almost precisely the same orbits, except for the orbit of Mercury. Mercury is the nearest planet to the sun, and it's affected more than the others by the sun's gravity (or, we could say, by the way the sun curves spacetime). Einstein's theory predicts a result of this proximity which is slightly different from the

The point of closest approach of Mercury to the Sun, the 'perihelion', advances by 1.38 arc seconds each orbit. The gravitational pull from the other planets, mostly from Venus, Earth and Jupiter, causes 1.28 arc seconds of this shift. General relativity explained the remaining 0.10-arc-second shift. (Figure not drawn to scale.)

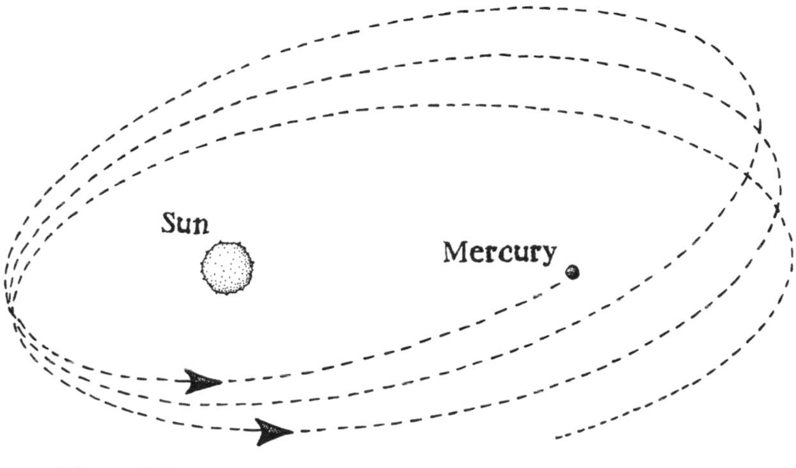

Figure 2.3.

result predicted by Newton's theory. Mercury's orbit as we observe it fits Einstein's prediction better than Newton's (Figure 2.3).

Interjecting a bit of science fiction trivia: nineteenth-century astronomers tried to explain the discrepancy between Mercury's orbit and Newton's theories by postulating the existence of an undiscovered planet whose gravitational pull affects Mercury's orbit. They christened the planet Vulcan. With the publication of Einstein's general theory of relativity, the existence of Vulcan was discredited (though some astronomers had claimed to have seen it). The writers of the television series *Star Trek* chose to immortalize that planet by giving the name Vulcan to the home planet of Spock.

We've said that even light cannot escape a black hole. We must now look again at that statement. It isn't clear in Newton's theories how gravity could affect light so that light itself would be trapped by a black hole. A bullet fired straight up from the surface of the earth will be slowed by gravity and eventually stop and fall back down. But what about a photon, a particle of light? The speed of light is fixed at 300,000 km per second. Even if escape velocity is greater than the speed of light, gravity (as Newton thought of it) can't slow down a photon and keep it from escaping. For the answer to this conundrum, we must look to Einstein.

3

The capture of light

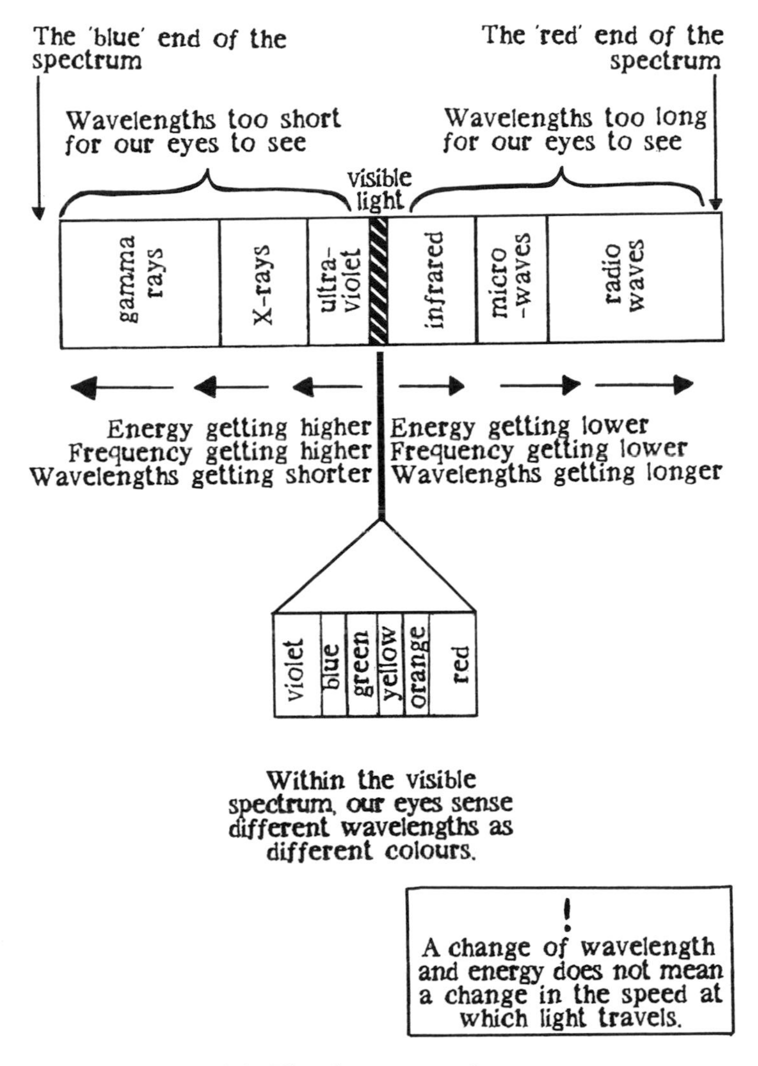

Figure 3.1. The electromagnetic spectrum.

When I tinker with my wireless set, I realize that all the sounds in the world are in my room.

George Bernard Shaw

When we say we are 'seeing', we mean that our eyes are receiving waves of light emitted from objects around us or bounced off them. We never see anything other than what arrives in our eyes by means of these waves. In fact, we don't see anywhere near everything that *does* arrive by means of these waves. We only see what comes in waves the right length to produce the colours of the rainbow or combinations of those colours. There are many other waves of light around us that we are never aware of at all. Perhaps it is confusing to call them all waves of 'light' when most of them are invisible. More correctly named, they are all waves within the electromagnetic spectrum. The part of that spectrum (a relatively small portion of it) that our eyes are able to receive is the 'visible spectrum' (Figure 3.1).

Radio waves are one category of electromagnetic waves beyond the visible spectrum. We recognize that radio waves are in the room with us only when we turn on a radio. The names of other parts of the electromagnetic spectrum are also fairly familiar: gamma rays, X-rays, ultraviolet rays, infrared rays, microwaves. In a later chapter we'll discuss the ways we are able to observe things in space that cannot be seen with optical telescopes by using telescopes that detect these other kinds of electromagnetic radiation. For now this mention of the spectrum is primarily a means of reminding us of a simple fact: if *no* electromagnetic waves in any part of the spectrum reach us from an object, not only is the object invisible to us, it is undetectable by all other instruments that detect electromagnetic radiation, such as radio or X-ray telescopes. If no electromagnetic radiation can escape from a black hole, it follows that none of the evidence we have of black holes can ever be in the form of any 'picture' of a black hole. All evidence we have is circumstantial evidence – what the black hole appears to be doing to objects and spacetime around it.

Should anyone be wondering why we spoke of light as 'particles' (photons) at the end of Chapter 2 and are now talking about it as 'waves', let us address that problem before proceeding. When we experiment with the way light propagates (the way it travels), we find that it behaves as though it were *waves*. The description of it as particles is ruled out. However, when we observe the way light interacts with matter, we discover that it behaves as though it must be *particles*. The model that describes it as waves is ruled out. By 1920, physicists had come to the conclusion that light could be thought of either in terms of waves or particles, but that neither model alone was adequate to explain the experimental data. This strange situation did not seem resolvable by saying that light is sometimes particles and sometimes waves. For a while, there were attempts to play one description off against the other and to decide that one is correct and the other incorrect, but it was soon frustratingly clear that the most practical way to proceed was to accept the two descriptions as incompatible but both necessary. Physicists found that the problem occurs with matter as well as with radiation. The description of an electron as a particle of matter cannot account for all the data. There are instances in which that model is ruled out.

Although many persons three-quarters of a century later still think of 'wave-particle duality' as a 'problem' in physics, the fact is that the mathematical physicist Paul Dirac soon discovered a theory which succeeds in combining wave and particle in a description with no contradiction or paradox. In the underlying mathematics there is a flexibility that reflects the dualism we have been discussing, but a simple mathematical transformation is all that is required to rewrite the equations of motion (that have to do with particles) as a wave equation. This did not keep Einstein from saying, somewhat later, 'All these fifty years of pondering have not brought me any closer to answering the question, What are light quanta? Nowadays every Tom, Dick and Harry thinks he knows it, but he is mistaken.'

At the end of Chapter 2 we introduced the question of how a black hole can manage to keep radiation (in any part of the electromagnetic spectrum) from escaping, without the speed of light changing. We'll return to that now.

Einstein's general theory of relativity does not predict that only planets and stars are affected by the warp of spacetime. Elementary particles such as photons (particles of light and all other forms of electromagnetic radiation) also travel warped paths. Consider a beam of light from a distant star travelling past the sun and reaching the earth. When the beam of light passes close to the sun, the effect is something like what happens when the golf ball rolls across the trampoline past the bowling ball. The warping of spacetime near the sun causes the photons' path to bend slightly in the direction of the sun (Figure 3.2). The sun is too bright for us to see such starlight in everyday circumstances. However, during an eclipse it is possible to get a false impression about which direction such a beam of light is coming from and what the distant star's position is in the sky, if we don't take into account the way the sun is bending the path of a star's light. Our eyes and brains do not make corrections for any change in photons' direction of travel on the way from their source to us. We judge an object to be located in the direction from which the photons from it enter our eyes, which is why we can be fooled by a mirror. In the chapters about observational evidence for black holes, we shall see how astronomers make use of the fact that paths of light are bent by massive objects. One way to measure the mass of objects in space, even objects which cannot otherwise be detected, is by studying the way they bend the paths of light from more distance sources.

A black hole is an area where the warping of spacetime is far more severe than near our sun. As the Cambridge physicist Stephen Hawking describes it to his lecture audiences, if we are using the trampoline analogy, we don't just have a depression, we have a bottomless hole in the elastic sheet.

However, we must now try to imagine this happening in more dimensions than it can on a trampoline.

Call to mind the fantasy of the shrinking earth. When we hovered in our spacecraft at the place in space where the original surface had been, the effect of gravity there didn't change no matter how much the earth shrank. But meanwhile the gravity on the *surface* of the earth did increase as the earth shrank. The same can

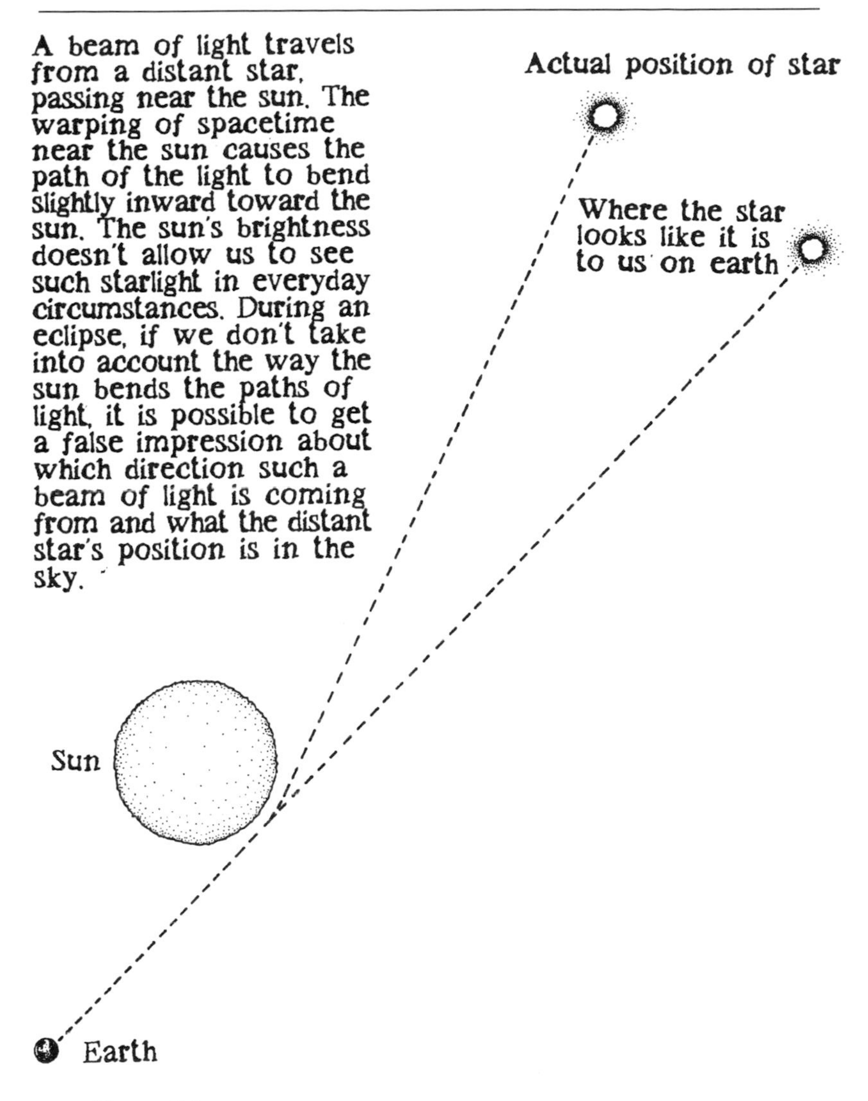

Figure 3.2.

be expected to happen on the surface of a collapsing star. Let us briefly focus our attention on the moment the collapsing star swallows its 'picture' in space and becomes a black hole.

The mass of the star is getting compacted into a smaller and smaller area, to greater and greater density. On the star's surface, the pull of gravity is growing increasingly strong. We find that the paths of light coming from other stars (light which would have hit the shrinking star's surface when that surface was larger) are more and more drastically bent near the star (see Figure 6.2 in Chapter 6). Meanwhile, light coming from the star itself finds it increasingly difficult to avoid following a path that curves back in.

Imagine a split second when the curvature of spacetime at the surface of the star is almost but not quite severe enough to bend the paths of photons coming from the star all the way back in on themselves. If we want to put that in terms of 'escape velocity', photons travel at lightspeed, and the escape velocity from the star's surface is not quite that great yet, but almost. We see the last photons that will ever escape from this star escaping in this split second, breaking for freedom with the beast of gravity nipping at their heels.

Imagine another split second just an instant later. The surface has shrunk a little further. Spacetime curvature at the surface is now severe enough to end all possibility of escape. The paths of photons coming from that surface are bent back in on themselves. We know that escape velocity from that surface is greater than the speed of light.

Now, getting to the crux of the matter: imagine an instant in time just *between* the two you have previously imagined (between the instant-of-last-escape and the instant-of-having-to-get-pulled-in). It is in that instant that the star officially becomes a black hole. For that brief instant, spacetime curvature on the surface becomes just sufficient to keep photons from escaping, but not sufficient to bend their paths all the way back in. We will continue to imagine a surface at this circumference, while the real surface of the star goes on collapsing, in the same way we continued to imagine a surface at the circumference in space where the original earth's surface used to be in the earth-shrinking fantasy. With a black hole, this imaginary surface, this 'boundary' where the black hole begins, is the 'event horizon'. The area inside it is the black hole. The area outside it is not. The event horizon is where the paths of light just fail to escape from the black hole but are not bent all the way back in. The photons trapped in these paths stay at that circumference as the star goes on shrinking. They won't get away (as did those that came before the surface had shrunk quite so far). Neither will they be pulled back in (as were those that were emitted when the surface had shrunk further). We can think of them as swarming in a thin, spherical shell, like a membrane surrounding the interior of the black hole. The star goes on shrinking inside this shell (Figure 3.3).

This is not easy to visualize. It helps to keep thinking of the story about the imaginary earth getting squeezed. There was a circumference in space where the surface had been before the squeezing. There the pull of gravity always felt exactly the same, even when the earth became a black hole. Similarly, the pull of gravity (the curvature of spacetime) at the circumference of the event horizon remains the same, even when the star has shrunk much further. That pull is exactly the right strength to keep the photons from escaping and at the same time not to pull them in. If they were the tiniest bit closer in, their paths would be so bent that they would curve back into the star. If they were the tiniest bit further out, they could follow a path that would lead to freedom. Where they are, they can, effectively, only spin their wheels. However, this photon shell is not visible from a distance as an orb shimmering in space. The photons cannot escape this circumference, and hence they can't reach our eyes. We can't see an object unless photons from it reach our eyes.

Unless the mass of the black hole changes, the event horizon will always stay where it was when the black hole first formed. Regardless of how much the star collapses inside it, the curvature of spacetime at this circumference will not change. We are going to learn that it is highly likely that more matter *will* fall into a black hole, which of course means it becomes more massive. Remembering Newton's laws, which tell us that an increase in mass means an increase in gravitational attraction, we know that 'all the way out', the pull of gravity (or spacetime curvature) will be greater than it was before. Any photons hovering at the original event horizon will be pulled in. There will be a new event horizon, its circumference larger than the old one. The black hole will have got slightly more obese.

Readers should be aware that there is a more detailed definition of the event horizon than we have used here, a definition

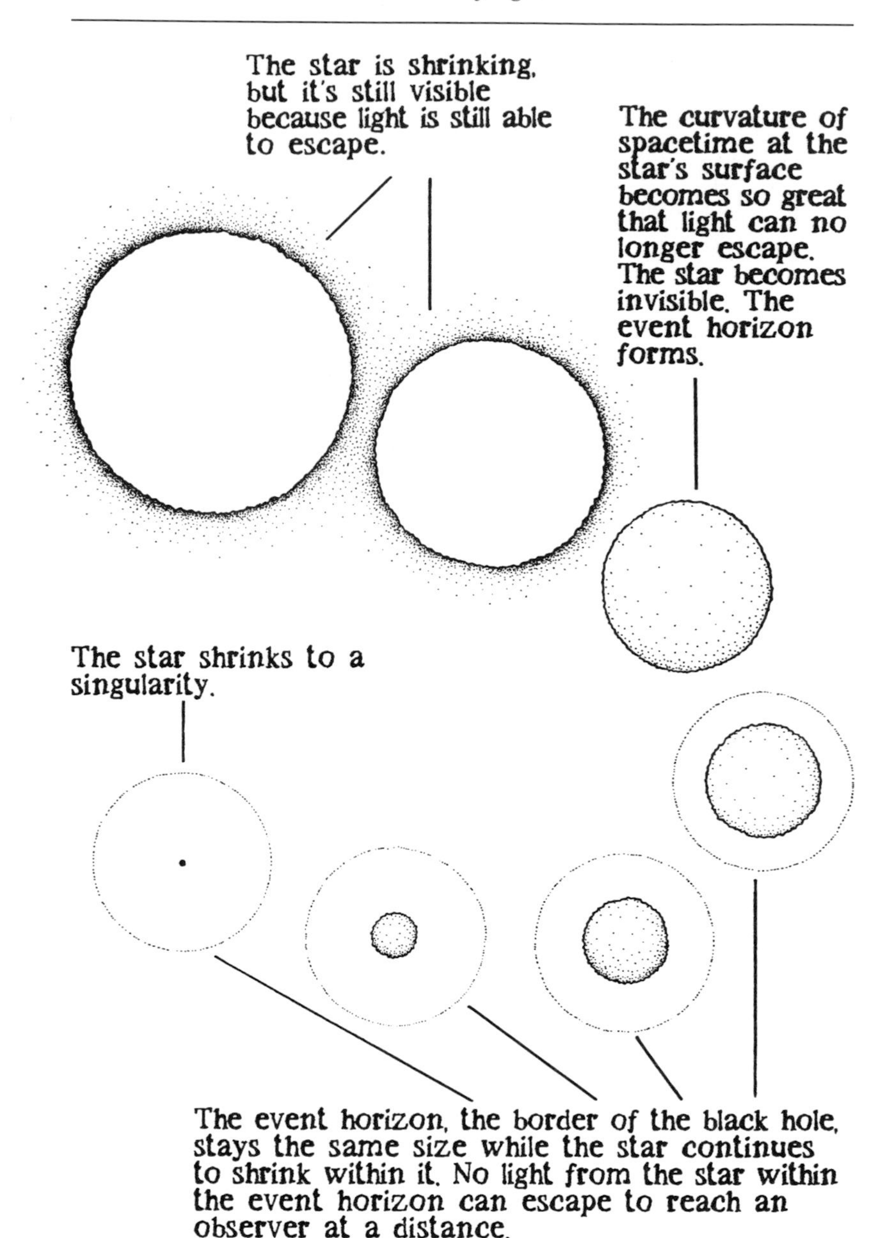

Figure 3.3. A star collapses and becomes a black hole.

introduced by Stephen Hawking in 1970. Knowing the distinction between these definitions will not be necessary in order to understand this book. However, for the curious, here it is. Before the star shrinks enough to prohibit the escape of light from its surface, the 'absolute horizon' forms first as a point in the centre of the star and then swells until it emerges through the star's surface just as that surface shrinks to the circumference we have been calling the 'event horizon'. After that, the two definitions are the same. In other words, at a time earlier in the collapse, when light can still escape to the distant universe from the star's surface, no light can escape from this steadily swelling region inside the star, a region that is *already* a black hole, a region whose border is the absolute horizon (see Figure 5.1 in Chapter 5).

The size of a black hole depends on its mass. In order to figure the circumference of a black hole – the circumference at which the event horizon (by our simpler definition) forms – take the solar mass of the black hole and multiply by 18.5. That will give you the circumference in kilometres. A 100-solar-mass black hole has a circumference of 1850 km. Using the same method, if we know the mass of a star, it is possible to predict what size black hole it will form – that is, unless the star loses weight in the process of collapse.

Can it ever lose weight *after* it becomes a black hole? How could this happen, if nothing can escape? We will hold this question for later.

Why should we concern ourselves so greatly about whether or not light can escape? It might seem more relevant to ask whether you or I could escape if our spacecraft slipped into a black hole. For something attempting to escape the earth's gravity, escape velocity is approximately 11 km (7 miles) per second. For anything attempting to get away from the event horizon of a black hole, escape velocity is just a little greater than the speed of light. Photons travelling at lightspeed are held captive there. According to Einstein's theories nothing moves faster than lightspeed. So nothing can outrun the gravity of a black hole. Cross the event horizon and we will not be seeing you again. You will be unable to report to the rest of us what it's like inside. From a black hole of the sort we've been talking about, no radiation of any kind (radio, microwave, X-ray, or any other), no sound, no sight, no space probe, absolutely no information can escape. Stephen Hawking and Oxford physicist Roger Penrose, in the late 1960s, suggested defining a black hole as an area of the universe, or a 'set of events', from which it is impossible for anything to escape to a distance. Theirs is now the accepted definition.

Returning to the fate of the star: even if it stopped shrinking immediately within the event horizon, we would still have a black hole. But it doesn't stop there. Having drawn the curtain, in complete privacy it goes on crunching down. Any light it emits, any portrait of itself that otherwise would be viewed from elsewhere in the universe, is pulled back in. It was Penrose who discovered in 1964 that a star collapsing as we have been describing has all its matter trapped inside its own surface by the force of its own gravity. Even if the collapse isn't perfectly spherical and smooth, the star goes on collapsing, with all the matter still trapped inside. According to general relativity, the enormous star is eventually confined not just in the region bounded by its event horizon but in a region that is nothing but an unimaginably small point. Here, everything that once was the star is concentrated to infinite density . . . to a 'singularity'. At a singularity, the curvature of spacetime is infinite.

Until the mid-1960s, though everyone had come to agree that general relativity predicts the existence of singularities, few took this prediction seriously. Physicists thought that a star of great enough mass *might* collapse to form a singularity. Penrose showed that if the universe obeys general relativity, it *must*. Trying to imagine what it would be like at a singularity is a nearly hopeless endeavour. Infinite density, infinite curvature of spacetime, end of space and time. Even today, more than thirty years after Penrose's discovery, our understanding of the laws of physics fails us here, and the best scientific minds in the world are not far better off than you and I. We're all a little out of our depth.

When a physicist finds infinities in his or her equations there is no way to proceed. We say that the theory breaks down. Hence, at a singularity the general theory of relativity breaks down. We reach the end of space and time as we understand them, and we do not yet know what other laws take over. For this reason, anticipating the new physics we expect to come into play at this juncture – the physics of 'quantum gravity', about which more later – it is customary these days to hedge our bets and speak not of a singularity of infinite density, but of 'near infinite density'. Our definition of 'singularity', then, is a place in spacetime where spacetime curvature becomes so strong that the laws of general relativity (which have led us there) break down and the laws of quantum gravity (which we don't know) take over.

Let us pause to sum up the main points of this chapter before proceeding. A star has collapsed and formed a black hole. We should be visualizing the event horizon as an invisible outer membrane or shell of the black hole – the outside edge of a spherical part of space with the singularity dead centre within it. We can think of the event horizon as an 'imaginary surface' in the same way we think of the equator as an 'imaginary line'. According to the theory we have looked at so far in this book, it is impossible for anything that can't go faster than the speed of light to escape from inside this surface and get away to a distance in space: no rocket ship, no space probe, no astronaut, no radio signal, no light – nothing at all. Photons at the event horizon can't be pulled in and can't get away. They just hover there. The singularity is an unimaginably small point at the exact centre of the black hole. Here, all the mass of the collapsing star has been compressed to near infinite density. The curvature of spacetime here is near infinite. Anything falling into the black hole will be drawn to the singularity. When it arrives there, it will have reached the end of space and time as we presently understand them.

Some of us would very much like to know what it's like in there between the event horizon and the singularity, and at the singularity. We can theorize about it, but we have little hope of collecting any direct evidence short of making a personal one-way journey. By comparison, just outside the event horizon the laws of physics as we know them are stretched almost to the breaking point but they still hold. We can make some predictions about what it would be like to visit there.

4

Tripping the theoretical fantastic

When I examine myself and my methods of thought,
I come close to the conclusion that the gift of fantasy has meant
more to me than my talent for absorbing positive knowledge.

Albert Einstein

If you have a yearning to 'go boldly where no man or woman has gone before', a black hole is surely the destination of choice, even if you venture only as far as the event horizon and are not so brash as to explore beyond that. Such an expedition is impossible with our present technology and with any we are likely to have soon. However, incredible voyages of discovery can be undertaken in our minds, and they needn't be strictly fantasy. It is possible to say what conditions near a black hole would be, based on what those who study black holes know or think they know about them at present. Well-established theory tells us a great deal about what we would encounter there.

On such an adventure we travel in good company. As Einstein indicated in the quote at the head of this chapter, journeys of this sort, albeit on a somewhat grander and more mathematical level than what we are about to undertake, are part of what theoretical physicists do for a living.

Let us set out then, bearing in mind four points.

1. We won't insist on knowing how we get to the black hole or how our ship is designed to survive the turbulence near a collapsing star. The design of such a ship must be left to the technology of the future. We shall talk about some of the problems but not expect to solve them here.
2. We won't insist on knowing how we could predict the collapse of a star with the sort of accuracy necessary to time our voyage properly, or how we could be sure that this particular star will not undergo a supernova which no ship in the vicinity could survive.
3. We must continually remind ourselves that it is a risky

business to take theory and translate it into 'what we would see' or 'what we would feel'. There are far too many ifs, ands, and buts about it. We'll do our best, but readers who are tempted to take this exercise much further on their own are warned that in order to do so they will have to know more about physics than this book can tell them.

4. The collapse as we are about to 'view' it is in many ways an idealization, simplified in order to increase our understanding of the concepts involved.

Our journey is to a star whose mass is approximately one hundred times the mass of our home sun. Experts on earth and aboard ship assure us that this star is on the brink of gravitational collapse. We all know from previous briefings that 100 solar masses is well above the Chandrasekhar limit and the limit for neutron stars. If black hole theory is correct, we are about to witness the birth of a black hole.

We must decide on a safe position for our ship, a distance from the star that will give us a good view but not put us too much at risk during the collapse. Also we must decide how closely we dare approach the black hole after it has formed. Much hangs on these calculations, because there is considerable potential danger, some of it unpredictable.

The hazard of tidal effects

Tidal effects are one of the easier black hole 'hazards' to understand, because we can connect them with something familiar, the ocean tides. Most of us have heard that the ocean tides are caused by the moon, but we may be rather vague about how the moon accomplishes this feat. We may have jumped to the conclusion that the moon's gravity is pulling directly on the ocean and drawing it upward in the direction of the moon, but there is a problem with that explanation.

Although it is the earth's gravity that keeps a NASA space shuttle orbiting the earth rather than haring off into outer space, astronauts aboard the shuttle don't feel any pull of the earth's gravity once they are in orbit. We have probably all seen pictures of them floating around the cabin of the shuttle. Their weightlessness is not caused by the spacecraft being too far away to be affected by the earth's gravity. It's due to the fact that the downward acceleration of the spacecraft exactly cancels the pull of that gravity. The astronauts, with their spacecraft, are in free fall, as surely as they would be if the shuttle were plummeting directly toward the earth. In free fall you don't feel any gravity. Remember that an orbiting spacecraft *is* falling, but its orbital motion causes it to fall *around* the earth, continually missing the earth, overshooting it again and again. That, roughly speaking, is what happens in any free fall orbit. A handy catch-phrase to keep in mind is 'Falling free or orbiting 'round, gravity cannot be found.' As we have said, acceleration exactly cancels out the earth's gravity, and the result is weightlessness.

Similarly, although it is the moon's and the earth's gravities that keep each of them in free fall orbit around the other (or, to be more precise, around their common centre of mass), we can't think of them or anything 'on board' either of them (such as the ocean) as directly experiencing the pull of the other's gravity, any more than astronauts in free fall orbit feel the earth's gravity. If this seems unlikely, recall that although it is the sun's gravity that causes the earth to orbit rather than fly off into outer space beyond the solar system, we don't weigh any less when the sun is directly overhead. Even on the planet Mercury, much closer to the sun, our weight wouldn't be affected by the sun being overhead.

Nevertheless, the ocean tides do occur, and we have already said that it's gravity that causes this tidal effect. What is this 'effect'? We find that the side of the earth closest to the moon (approximately 12,000 km nearer to the moon than the opposite side of the earth) bulges out toward the moon, and the side of the earth furthest from the moon bulges out away from the moon. The earth is stretched slightly, first in one direction and then in another, depending upon where the moon is in its orbit. The stretch doesn't occur only where there are great bodies of water, but those areas of the earth stretch more readily than other areas, and the effect there is noticeable. Whatever part of the ocean is closest to the moon bulges out in the direction of the moon, making a high tide.

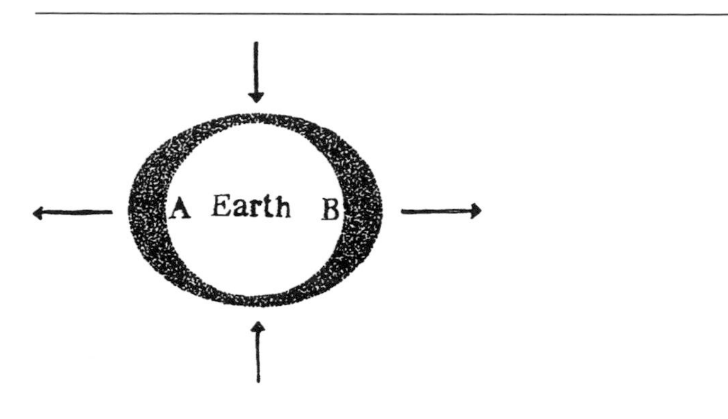

The shell of ocean around the earth bulges out toward the moon on the side nearest the moon (B), making a high tide, and bulges out away from the moon on the side farthest from the moon (A), making another high tide. Halfway in between we have low tides. It's the difference in the way the moon's gravity pulls on different parts of the earth, not the direct gravity itself, that causes this stretching and squeezing. See Figures 4.2 and 4.3 for an explanation of how tidal gravity works.

Figure 4.1. Tidal effects.

Whatever part of the ocean is furthest from moon bulges out in the direction away from the moon, making another high tide. Halfway in between we have low tides (Figure 4.1). We are right to think gravity is the culprit behind all this, and it isn't entirely accurate in our catch-phrase to say that 'gravity cannot be found'. In order to understand how the deed is done, you need to know something about 'tidal gravity'.

Let us imagine that from some point above the earth's surface we drop a large box toward the earth with two test objects in it. The entire contraption is in free fall. We know in principle that if we were in the box with the two objects, we would be aware of no pull from the earth's gravity. We would be weightless, and we would have no sense of what was 'up' and what was 'down'. Now, if there were no large concentration of mass such as the earth nearby, and if these two test objects were somewhere out in empty space alone, their mutual gravity would set them in free fall toward one another. But the earth's gravity is so enormously much greater

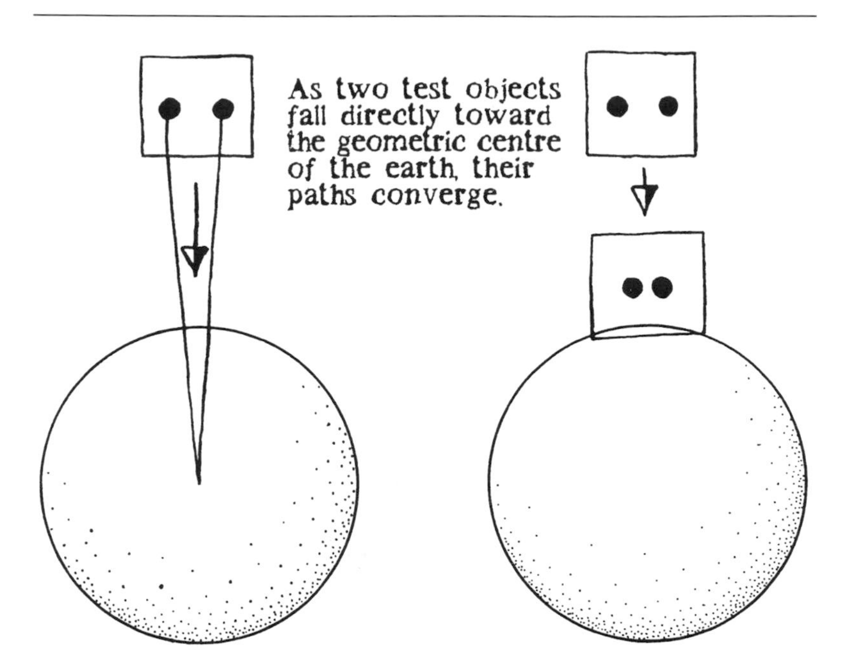

Figure 4.2. Two test objects fall toward the earth.

than that of these two test objects that we can safely ignore that effect as we proceed. Our two objects are each falling 'down' directly toward the geometric centre of the earth, and that means their paths as they fall are converging. Paths that begin separately have to converge if they are to arrive at precisely the same point. Hence, the two test objects get closer to one another as they fall (Figure 4.2). Although we, with them inside the falling box, feel weightless and would seem to have no way of knowing we are falling, we *can* tell we are falling toward a much larger object because of the fact that the two test objects are getting closer to one another. This indirect effect of gravity is a giveaway.

Let us imagine that instead of two test objects we have four. In addition to the two we had with us previously in the falling box, we have a third a little 'lower' (in relation to the earth) and a fourth a little 'higher' (Figure 4.3). The resulting arrangement of the objects is a square with one point of the square (one object) leading the way down as they fall and the opposite point (another of the

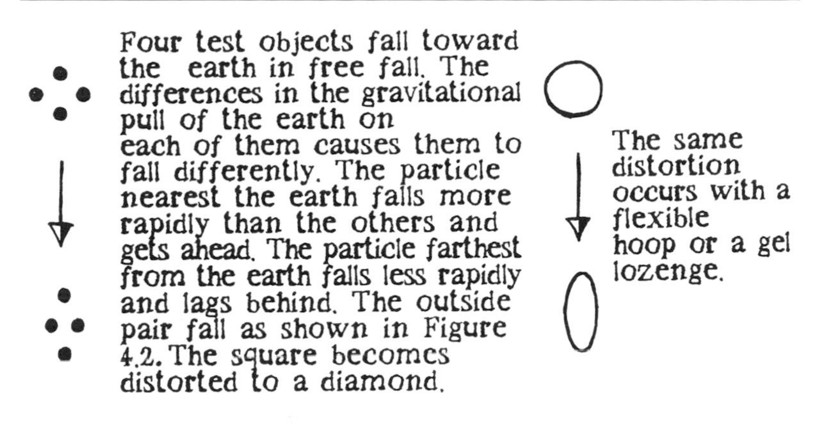

Figure 4.3. Four test objects and a lozenge fall toward the earth.

four objects) coming last. (Since this is difficult to describe in words, please refer to Figure 4.3.) All four objects are in free fall. Once again, in principle, if we were in the box with them we would not feel any pull of gravity or have any feeling of 'up' or 'down'. Nevertheless, again we could deduce that we were falling toward a much larger object because of an indirect effect of the earth's gravity: gravity would cause them to fall at slightly different speeds. The 'highest' object is the greatest distance from the earth. We learned in the discussion of Newton's laws that the pull of gravity decreases with distance, so the attraction between this highest object and the earth is less than the attraction between its three companions and the earth. It lags further and further behind the others as they fall. The 'lowest' object is closest to the earth. The attraction between it and the earth is greater than the attraction between its three companions and the earth. It increases its lead as they fall. The other two objects fall as they did before, with their paths converging. The square gets distorted. It is the *difference* in the strength of gravity from place to place that causes the distortion.

Understanding that, we can also understand the part of Figure 4.3 which shows the same effect as it happens with a flexible hoop or, in three dimensions, a gel lozenge falling toward the earth, and we can begin to see why the tides occur. This vertical stretch and lateral squeeze are called 'tidal gravitational forces' or 'tidal gravity'. Looking at the same stretch and squeeze through Einstein's eyes, curvature of spacetime and tidal gravity are the same thing.

The earth and the moon, in free fall orbit, are not falling directly toward one another as the lozenge is falling toward the earth in our example. But, just as surely as if they were, they are in free fall, and nothing on board the moon or earth can feel the direct gravity of the other. However, as happened with the gel lozenge, the earth, the moon, and things on board them can be affected by the *difference* in the strength of the gravitational pull from place to place and the *difference* that makes in the way they 'fall'. The upshot is that the side of the earth nearest the moon bulges out moonward, making a high tide. The opposite side gets 'left behind' a little, making a second high tide. In between there are low tides. If the entire earth's surface were as flexible as the gel lozenge, we would not have to go to the shore to experience the tides. To finish off this discussion we should note that the moon undergoes a similar stretching as an indirect effect of the earth's gravity. However, there are no bodies of water on the moon to make the effect obvious, and the moon is no gel lozenge.

If anyone needs further convincing of the difference between direct gravity and tidal gravity, note that although the sun's gravity at the surface of the earth is about 180 times stronger than the moon's gravity at the surface of the earth, solar tides are not as big as lunar tides. The difference between the pull of the sun's gravity on the side of the earth nearest the sun and on the side of the earth furthest from the sun is small. The sun is much further away from us than the moon, and the distance from one side of the earth to the other is so small compared to the earth's distance from the sun that that distance makes very little difference in the pull of the sun's gravity.

Though tides are known to cause problems for us living on the earth, most of us don't often have reason to consider tidal effects a major threat to our well-being. Travelling near a black hole, we are advised to treat them with far greater respect. From the above paragraphs, we can guess that it isn't the enormous strength of gravity that will present a hazard in this case, even *within* the black hole, but rather the extremely rapid *change* in the strength of gravity as one approaches or moves away from the event horizon. In our example, the 'highest' and 'lowest' of the falling objects fell at slightly different rates. Near a black hole these same two objects would fall at *vastly* different rates. If they are two parts of a lozenge, they will likewise fall at vastly different rates and the lozenge won't last long as a lozenge. If they are two parts of our ship – or of you or me – we are in trouble.

Gravitational attraction or spacetime curvature increases rapidly as one approaches the event horizon and falls off just as rapidly as one moves away from the hole. The change may be over short, even microscopic, distances. The effect on one side of the ship, the side facing the black hole, could differ drastically from the effect on the side away from the black hole. The closer the spacecraft approaches the black hole, the greater the difference in these effects on various parts of its structure will become. At some point the ship will not merely be stretched. The ship and its passengers will be ripped asunder. Imagine what would happen in Figure 4.3 if this were very near a black hole and the 'lowest' of the test objects were your feet, the 'highest' your head, those at the sides your left and right hip. Larger objects will feel the effect most dramatically at first, because their sides are further apart than those of smaller objects and the difference in the strengths of gravitational pull will be greater. However, near the singularity, the point at the centre of the black hole, even elementary particles will be unable to hold together.

Calculating a safe orbit or position to assume after the black hole forms involves more than simply figuring out where the event horizon will be, so that we won't slip beneath it and be unable to get back out again. We must take into account the tolerance of the ship and its passengers to various degrees of stretching and compressing. We must also consider the size of the black hole. If it is a small black hole, that means the event horizon will be near to the singularity, and even outside the event horizon the tidal effects will be enormous. If it is a large black hole, such as the extremely massive ones we think are at the centres of galaxies and quasars, the event horizon will be far from the singularity. In principle, we might well come to no harm at all from tidal effects from a hole of that size even travelling some distance below the event horizon, though we would still be unable to escape. Unfortunately, the 100-solar-mass black hole we will be visiting is a relatively small one. If we venture too near, the result will not be pretty. We might think it wise to position the ship, and ourselves in it, so that the part nearest the black hole is the least possible distance from the part furthest from the black hole. However, there are other hazards to be taken into account.

The hazard of gravity waves

Tidal effects are well known and well documented. 'Gravity waves', on the other hand, at the time of writing this book have not been directly observed in any of the experiments designed to detect them. However, Einstein's general theory of relativity predicts their existence and tells us that when a star collapses to form a black hole there is likely to be a burst of them.

A gravity wave shows itself through a travelling disturbance in spacetime curvature just as a wave in water shows itself through a travelling ripple in the surface of a pond. We can imagine dropping a stone into a pond and seeing ripples or waves from it move out in expanding circles of peaks and troughs. Around a collapsing star, gravity waves may move out in the form of expanding spherical zones (the equivalent of circles within circles, but in more dimensions) of negative and positive spacetime curvature, a series of pulses. Each bit of curvature moves outward at the speed of light.

In order to get a better grip on the concepts involved here, we'll once again imagine some test objects. As with the tidal effects we discussed above, looking at one small test object in free fall would tell us nothing. The effect can only be noticed if we watch two nearby objects and observe the change of position of one relative to the other. We will imagine placing the two small test objects side by side, at equal distances from the collapsing star (Figure 4.4). They are in space too far away from the star for its regular gravitational attraction to have any significant effect on them. A gravity wave approaches. If the first part of the pulse has negative

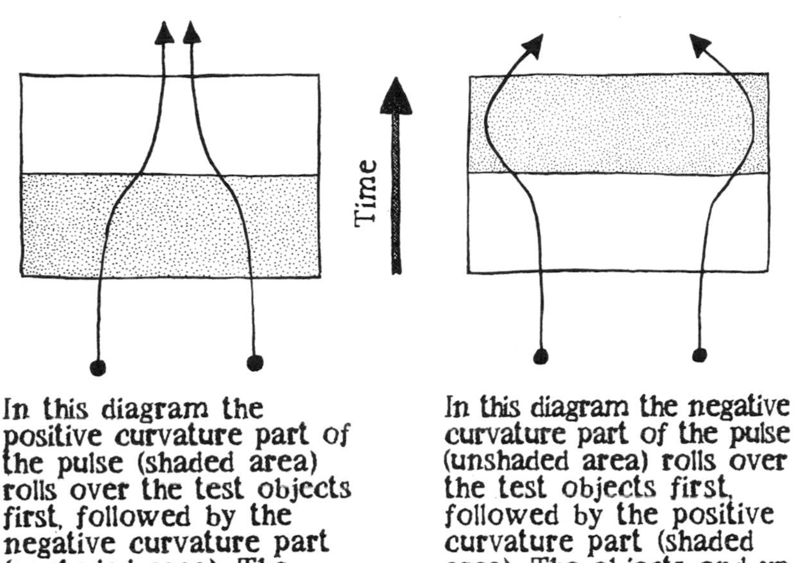

Figure 4.4. Test objects and a gravity wave.

spacetime curvature, as it rolls over the two test objects the separation between the two will suddenly grow greater. When the trailing, positive-curvature part of the pulse passes through, that will cause a counter-effect and leave the objects on paths that approach one another. Or it could happen the other way around, with a decrease in the objects' separation happening first. This second scenario would leave the objects travelling on parallel paths, but closer to one another than previously. However, if we start out with the two objects arranged one directly 'above' the other ('down' being in the direction of the source of the gravity wave), the separation between them will not change.

It should not be too difficult to see that these ripples of curvature in spacetime could put a considerable strain on the structure of our ship and on our bodies, alternately shrinking and stretching by the same process that causes the two side-by-side test objects to increase and decrease their separation, and thereby setting up what might prove to be intolerable stresses. The more smooth and near spherical the star and the star's collapse are, the less we have to worry about. If the star's distribution of mass is spherically symmetric (no lumps, bumps, bulges, or other departures of the star's mass from sphericity, anywhere) and stays spherically symmetric until the star shrinks below its event horizon, there will be no gravity waves at all. A steady change in the distribution of the star's mass will cause no gravity waves. It takes a whole string of abrupt and sporadic changes in the pattern of the collapsing star's mass to produce a train of waves. However, if we are dealing with a real-life star, not an ideal, it would be foolhardy to think there are no lumps and bumps – foolhardy to think there will not be a string of abrupt and sporadic changes in the distribution of mass as the star shrinks.

The experts on board our ship will already have done some calculations. We know, of course, that we will not feel the effect of a gravity wave directly as anything like an increase of weight of ourselves or our ship. We know the rate at which the strength of a gravity wave should decrease with distance from the source – a somewhat slower rate than that at which regular gravitational attraction decreases with distance. Best expert opinion tells us that the star is unlikely to radiate away more than a small fraction of its total mass/energy. However, no one knows how violent and frequent and sporadic the changes in the pattern and arrangement of the star's mass will be during the collapse. How dramatic the burst of gravity waves will be depends critically on details of this particular collapse – details that we have no way of predicting. It would be wise to prepare for the worst.

Earlier we talked about the potential hazard of tidal effects near a black hole and agreed to keep a judicious distance away. The hazard of gravity waves during the collapse will cause us to stay back much further. The rate at which the strength of a gravity wave decreases with distance from the source is, as mentioned above, a somewhat slower rate than that at which regular gravitational attraction decreases with distance.

One of our more optimistic passengers has suggested we should be encouraged by the fact that the collapse happens quickly (as time is measured on the surface of the star), and gravity waves cannot escape from within the event horizon, once it forms, any more than anything else can. A more sober assessment tells us that because of gravitational time dilation (which we'll learn more about later), the last gravity waves emitted by the collapsing star just before the star reaches the circumference of the event horizon will take a long time to get to us. We're going to be experiencing these things for a while! What is more, if the black hole later swallows anything very massive, there may be a considerable burp of gravity waves from just outside the event horizon. Never mind, says our optimist, those gravity waves will be 'redshifted' too far to do damage. But who says a redshifted gravity wave is any less destructive?

Our 'optimist' is a little ahead of us. Let us pause to find out what is meant by 'redshift'.

Gravitational redshift

One bit of science most of us learned as small children is the Doppler effect. It was my father who explained to me and my brother and sister, when a locomotive passed at high speed and we heard a dramatic drop in the pitch of its whistle, that the sound of the whistle came to our ears as waves like waves on the ocean, and as something moves away from us the sound waves coming from it to us get stretched. That is, the distances between the wave crests get longer. Our ears interpret longer sound waves as lower pitches.

The Doppler effect occurs with all types of waves – light waves, gravity waves, even waves on water. Let us first see how it happens with light.

We have said that our ears interpret longer sound waves as lower pitches. Similarly, our eyes interpret different lengths of light waves as different colours, the longer the light waves the nearer the 'red' end of the spectrum (refer to Figure 3.1 in Chapter 3). If an object continues to accelerate away from us, increasing its speed as it goes, the light waves coming from it to us are stretched –

'redshifted' – more and more, perhaps to such lengths that they move beyond the visible part of the spectrum. In that case, the object becomes invisible. Keep in mind, however, that only if the source of the light waves were retreating at a speed approaching the speed of light would any redshift be noticeable to our eyes. We need not look for it in everyday circumstances.

Light reaching our spacecraft from the collapsing star will be redshifted. Why? The star as a whole won't be accelerating away from us, but the surface of the star nearest to us will. That surface will be receding more and more rapidly as the collapse continues. Similarly, if you hold tightly onto a balloon and let the air out, the balloon may not move away from you, but the nearest surface of the balloon will.

What we have been talking about here is the way *acceleration* can stretch sound or light waves. One principle that Einstein brought to our attention is that gravity acts like acceleration, and vice versa. Gravity and acceleration can feel the same and have the same effect on things. The most familiar example has us taking an elevator from the ground floor of a building to the top floor. The elevator door closes. For a few seconds we feel heavier, as though gravity were pulling down harder on us than it normally does. In fact, the pull of the earth's gravity has not changed. We feel heavier because the elevator is accelerating, reaching higher and higher speeds as it begins its climb. Next, the elevator reaches its maximum speed and the speed becomes constant. Zero acceleration. We feel only the normal pull of the earth's gravity, holding us comfortably to the floor of the elevator. Then the elevator nears the top of the building and begins to slow down. We experience deceleration, or negative acceleration. Gravity seems to have become weaker. We feel lighter. We know very well that the pull of the earth's gravity has not changed while we have been on the elevator. The effects of acceleration and deceleration on our bodies fooled us into thinking it had.

Another example: in a spacecraft too far away from any astronomical body to be affected by its gravity, travelling at a constant speed in straight line motion, passengers float free with no sensation of gravity. If the spacecraft accelerates, they feel as though gravity has been switched on. If the spacecraft continues to increase its speed at a steady rate of increase, the effect may be indistinguishable from what they would feel living in a gravitational field. Depending upon the rate of increase, they may feel heavier or lighter, but one surface of the passenger compartment, the surface in the opposite direction from the direction of acceleration, will become for them 'the floor'. Everything on board ship – furniture, liquids, golf balls, elementary particles – will behave as though that were indeed the floor. In a spacecraft with no windows, no engine noise or vibrations, with a rate of acceleration just right to simulate the earth's gravity, the passengers would not know whether they were in outer space or still waiting on the launch pad on earth.

The equivalence of gravity and acceleration that we have been demonstrating means that gravity also stretches waves. We call the result *gravitational redshift*. There will be more about gravitational redshift in later chapters, but just now we must return to the subject that required this digression in the first place – gravity waves and how they might affect our ship. The best route back into that discussion is first to see how gravity affects light waves coming from the star's surface.

We can predict that light waves coming to our ship from the collapsing star will be stretched toward the red end of the spectrum not only by the surface's acceleration away from us but also by the increasing gravitational pull at the star's surface. Even before the star collapses enough to form the event horizon, the light will have redshifted out of the visible spectrum and even beyond the radio spectrum. From our ship, we won't see the star 'wink out' as the event horizon forms. It will become invisible to us before that. Eventually the wavelengths will become too long for us to detect with any equipment. At the event horizon, the redshift becomes infinite.

Another way of thinking about this is to say that the light uses up all its energy getting away from the edge of the black hole. Figure 3.1 showed that lower energy levels go along with movement toward the red end of the spectrum. Hence, photons from near the event horizon reach us, arriving at the speed of light, but with no energy left at all.

That, broadly speaking, is what happens to light waves coming to us from the star. What about gravity waves? The expectation is that they also will be redshifted, but does that mean they will not be a hazard to us? We can't be sure. We do know that no gravity waves can reach us from within the event horizon, though this should not mislead us into thinking that the mass of the black hole will have no gravitational effect. It will continue to warp spacetime, affecting other objects such as our ship or nearby astronomical bodies.

Hazards near a rotating black hole

It is likely that the star we're watching, which is about to become a black hole, is rotating. If so, we can expect the rate of its rotation to increase enormously as the star crunches down and becomes a black hole. Why should the rate increase so much as the star collapses?

We can demonstrate the same phenomenon on a children's merry-go-round. If all the children perch on the outer edge and lean as far out as possible, the merry-go-round slows down and almost comes to a stop. If all the children move toward the centre of the merry-go-round, it speeds up.

Or . . . consider a skater doing a spin. With one leg and both arms extended, she spins rather slowly on one skate. When she draws her leg and arms tightly in against her body, the rate of spin increases until she is a blur.

Or . . . consider a planet orbiting the sun in an eliptical orbit. The planet speeds up in the part of its orbit during which it is approaching the sun, and it slows down in the part of its orbit during which it is moving away from the sun.

In all four examples we seem to be getting something for nothing. Nothing comes along to spin up the star as it collapses. No child hops off the merry-go-round to give it a push. The skater doesn't reach down with her free skate to give an extra kick against the ice. There are no rockets attached to the planet. But none of these are instances of getting something for nothing. They are cases of keeping something that is already there. What is kept, or 'conserved', is the 'angular momentum'. There is a law of nature which requires that angular momentum be conserved, which means that though it can be transferred from one object to another, angular momentum cannot simply appear out of nowhere, nor can it disappear. Hence, when the angular momentum isn't needed to pull around things that are far out from the axis of a spinning object, as the children are when they lean far out from the merry-go-round, it has to show up in another way. As the children move toward the axis of rotation, the angular momentum goes into speeding up the rate of rotation. When we see the skater draw in her arms and leg, concentrating her mass closer to her axis of rotation, we know that angular momentum is conserved by the resulting increase in her rotation rate.

The collapse of a star to become a black hole is a prodigious drawing in of arms and legs. The black hole will have all the angular momentum the star had, but the mass will be concentrated nearer and nearer the axis of rotation. If the star was spinning at all, the black hole will have an enormous rate of rotation.

Although it would be misleading to think of spacetime as a substance like water, we can to a certain extent demonstrate the odd effects a rotating body has on things in spacetime around it by spinning a ball in water. Objects floating near the ball tend to get dragged along by its rotation. The faster the ball spins the stronger the effect.

Objects in the vicinity of a rotating black hole feel a similar effect in a powerful way, increasingly powerful the nearer they approach the event horizon. Come close enough to the event horizon (still outside it) and the dragging becomes so powerful that nothing can remain at rest. Everything there is swept around in the same direction the black hole is rotating. If our ship were to enter that area, we could blast our rocket engines as hard as possible and it wouldn't do us a bit of good, we would not be able to resist being swept along.

The 'ergosphere' is the name given to the area around a rotating black hole where the dragging is too strong to resist. The outer border of the ergosphere is the 'static limit', or the 'stationary limit'. Inside this limit nothing can remain stationary (Figure 4.5).

Outside the static limit we can maintain a position that is stationary in relation to distant stars – a position from which we continue

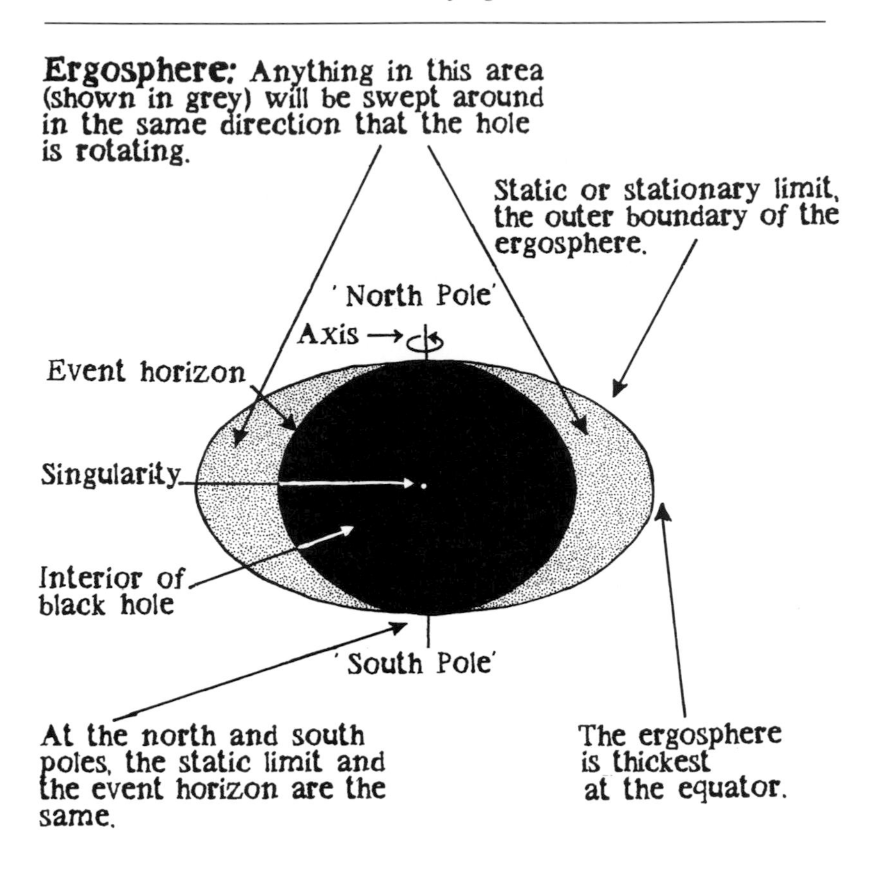

Figure 4.5. The ergosphere.

to see the same pattern of stars while the black hole rotates beneath us. All that is required is a sufficiently powerful downward blast of our rocket engines. Beneath the static limit, within the ergosphere, no conceivable rocket engine could allow us to maintain such a position. We might rev up the engines enough (still directing them downward) to allow us to keep a fixed distance from the event horizon of the black hole and avoid being dragged closer, but we could not avoid being dragged along in the direction of the rotation of the black hole. Travelling in a rocket-supported orbit not too near the event horizon, we would be able to change somewhat the rate at which we went around. But that rate could never be zero. Go we would. Nearer the horizon we could change hardly at all the rate at which we went around.

We may attempt some adventures in the ergosphere when we've become more familiar with conditions around this particular black hole.

5

Crossing the bar

And ere a man hath power to say, 'Behold!'
The jaws of darkness do devour it up.
William Shakespeare, *A Midsummer Night's Dream*

As a way of previewing what will happen as the star collapses, we should all familiarize ourselves with Figure 5.1. This is a diagram with a time line. Those who are unaccustomed to such diagrams would do well to study this one with special care.

The time line is the arrow on the left side, appropriately labelled 'time'. It serves as a clock for the diagram, indicating that if we move our eyes up the page we move forward in time. For example, look at the left hand page of the figure. Beginning at the bottom, notice that the centre area represents the star. Its collapse hasn't yet begun. Move up the page and we find that that area gets smaller, which shows us that the star is collapsing as time progresses. When the collapse has barely begun, a light ray leaves the star and heads off into space (on the right-hand side of the diagram). A little above that (a little later, as the time-line indicates), the event horizon forms – a shaded area that doesn't get smaller, as the star, shown now in shadow, continues to collapse within it. We see a light ray staying at the event horizon as time passes. This is one of the light rays that is emitted at the instant the event horizon forms. Unless the mass of the black hole changes, that light ray will never escape nor will it be drawn in. The fact that the diagram shows the star in shadow after the event horizon forms reminds us that no light is escaping to reach distant

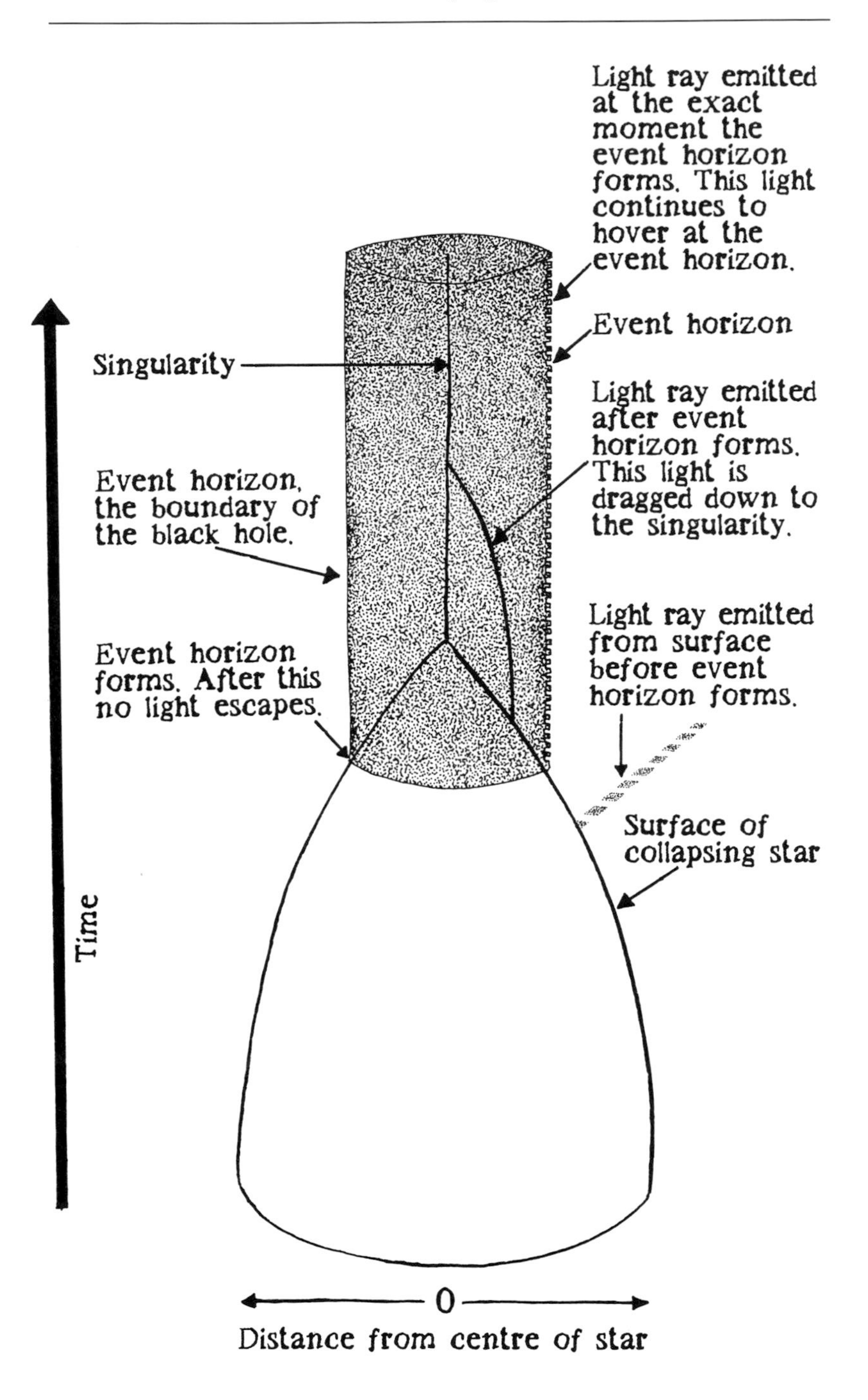
Light ray emitted at the exact moment the event horizon forms. This light continues to hover at the event horizon.
Event horizon
Singularity
Light ray emitted after event horizon forms. This light is dragged down to the singularity.
Event horizon, the boundary of the black hole.
Light ray emitted from surface before event horizon forms.
Event horizon forms. After this no light escapes.
Surface of collapsing star
Time
0
Distance from centre of star

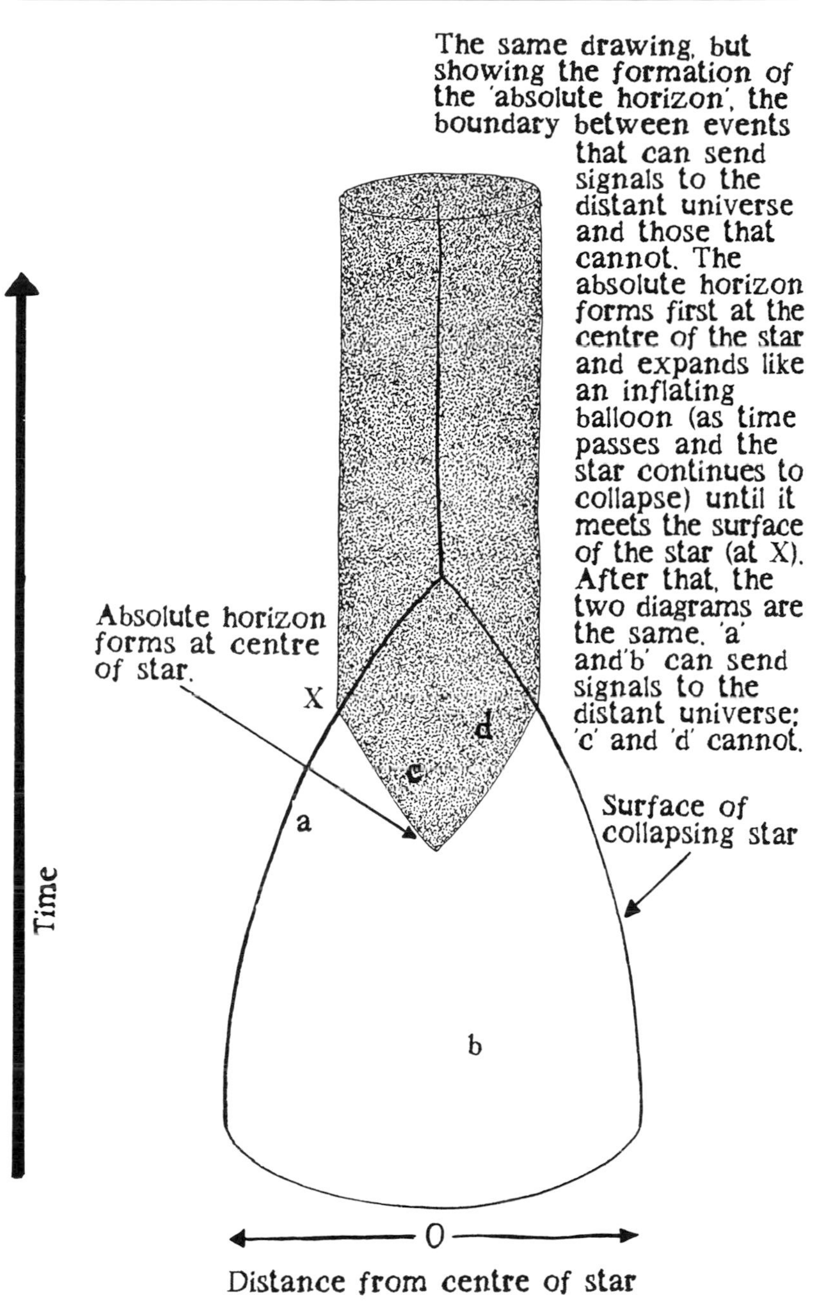

Figure 5.1. A star collapses to form a black hole and a singularity.

observers. Time marches on (up) and the star crunches down to a point – a singularity. The diagram shows a light ray being dragged to the singularity.

At the bottom of the diagram we have a star. Before the time-line reaches the top of the diagram we have a black hole. The right hand page of the figure shows the same sequence using the concept of the 'absolute horizon', which we spoke of in Chapter 3.

In the crunch

In books such as this it is obligatory that an intrepid astronaut go down into the black hole. We will not neglect this requirement, but we'll do it a little differently. We'll send the astronaut down to the surface of the star early, before the event horizon forms, so that he or she can 'ride' that surface while the star crunches down. A communication device built into the astronaut's space suit will send a beep and a flash of light every second, as measured by an accurate clock attached to the device. We need a volunteer.

With the volunteer (of course there is one, human curiosity being insatiable) in place, we are ready to observe the collapse. In any real collapse of a star, dust and debris would almost surely obscure the view from the ship, and no astronaut could stand on the surface of a star. However, we are viewing all this with the eyes of theory and imagination, so for the sake of learning about black holes let us suppose that we can see the star and that the astronaut can assume a position there. We will do two 'takes' on this collapse, one from the point of view of those of us who have been wise enough to remain on the ship and a second from the point of view of the foolhardy volunteer down on the star's surface.

From our point of view

We are gathered on the bridge of the ship. The star is visible on the viewing screen, magnified somewhat by the ship's instruments, because from this distance – a safe distance, our experts think –

the last stages of the collapse will not be easy to observe. The beeps from the astronaut's communicator and the flashes of light come at regular one-second intervals.

The ship shudders and we sense a change in the deep rumble of the engines. Engineering is repositioning the ship as experts have directed, to take the least possible damage from gravity waves.

At three fifty-nine and one second, we receive the first of the final sixty signals that will bring us to four o'clock. The volunteer, or perhaps only the communicator and clock, is evidently still alive on the surface of the star. We count down the seconds approaching four o'clock, one beep and flash every second, regularly. At first there is no apparent change in the star or the intervals between the signals. Then, sure enough, the star begins to shrink into itself faster and faster like water going down a drain. An ominous groan from the ship . . . a snapping sensation in our bodies. The shrinking stops and the star hangs frozen in space, then turns red. We don't need our ship's sensitive instruments to tell us the fifty-ninth signal is late, and this time the flash of light is reddened. The star darkens, becomes only a shadow and then completely black. Nothing. Our eyes and ears still await the next signal, but there is none. We will never hear or see the four-o'clock signal.

From the astronaut's point of view

The intrepid volunteer has landed on the star and flicked the switch on the communicator. It emits signals at regular one-second intervals. Soon the star begins to shrink. How does the astronaut know? Does it feel like an earthquake? At some point, perhaps so, when matter becomes compressed to the density of a solid. There is a noticeable redshift of distant stars and galaxies as the astronaut, riding the surface of the collapsing star, accelerates away from them. There are weird effects as beams of this light are drastically bent by the increasing gravitational pull at the star's surface. The fifty-ninth signal is not late. Neither is the sixtieth, as the star shrinks beyond the event horizon with the astronaut (assuming he or she might have survived the tidal effects so long) having no way of knowing this is happening. A fraction of a second later the astronaut surely has no way of knowing *anything* is happening, as astronaut and communicator are ripped apart by tidal effects and fall to the singularity.

There are puzzles here. Why was the fifty-ninth signal late to our ears and eyes but not to the astronaut's? Why was the fifty-ninth flash reddened? Why did the astronaut hear and see the sixtieth signal while we did not? Why did the star's image freeze and then fade rather than shrink steadily to nothing? What has happened to the astronaut?

First, the question of the signals . . . and the first of several impossible things to believe before breakfast.

Time marches out of step

In old adventure and war films, we hear the command given, 'Synchronize your watches.' In inter-galactic wars, obedience to that command might not help anyone much, because time doesn't pass at the same rate everywhere in the universe, and watches quickly go out of synchrony.

Wherever you or I might happen to be in the universe, time there would seem to us to pass in its familiar manner. As far as we know, there is no area where the inhabitants experience time passing as it does in fast-forward on a video, or in slow-motion. However, if one observer were able to be simultaneously in two different locations in the universe, that observer could discover mind-boggling differences in the rate at which time passes.

This bizarre effect is 'time dilation', and it allows us to make two statements which contradict one another and nevertheless are both true:

Statement 1: 'As the star collapses, the surface of the star, the astronaut, and the communicator take an infinite amount of time to cross the event horizon. Time slows down to a complete stop there. Star surface, astronaut, and communicator are stuck there for eternity. They never reach a singularity.'

We can make that statement and be quite correct in making it if we are observers at a distance from the black hole. From our vantage point, the dilation of time at the event horizon becomes infinite, and that means that by our clocks out here, time down there at the event horizon does stop.

Statement 2: 'As the star collapses, the surface of the star, the astronaut, and the communicator cross the event horizon and reach a singularity very quickly. They must arrive there just as surely as we all must arrive at a point in time that we call tonight at 11 p.m. It can't be avoided.'

We can make that statement and be quite correct in making it if we are the astronaut on the surface of the star. From the astronaut's point of view, the journey across the event horizon and to the singularity takes place very rapidly indeed.

Obviously we have a discrepancy, if not complete nonsense. But neither statement is wrong. It all depends on where you are.

One key to understanding time dilation is to realize that redshift and time dilation are two faces of the same coin. We saw in Chapter 4 that gravity and acceleration can have almost identical effects. Both can stretch electromagnetic waves – in fact, waves of all kinds – to greater wavelengths. Likewise, they can stretch the interval between pulses. If the astronaut, near the event horizon, sent waves or pulses of light or beeps or anything else at regularly spaced intervals, those waves or pulses or beeps would arrive at our ship less frequently (a longer interval between waves or pulses or beeps) than the astronaut sent them.

Let us try to get a grip on this by reviewing some of what we've discussed before. Waves of visible light are electromagnetic waves. The longer the wavelengths, the further they are toward the 'red' end of the spectrum (review Figure 3.1). Hence, to stretch wavelengths is to 'redshift' them, and this term is used whether they are visible light or any other type of electromagnetic radiation. We have said that light waves coming to our ship from the shrinking star would be stretched by both the acceleration of the star's surface as it recedes and the rapid increase of gravitational attraction on the star's surface.

Saying that wavelengths are stretched is the same as saying that the crests of the waves reach us less frequently – at a slowed down rate. We can easily visualize this as the crests of ocean waves as they reach us on the beach. 'Frequency' is a term we also are familiar with if we use a radio. The term refers to how frequently the wave crests arrive. If they arrive more frequently (shorter wavelengths), we say they have a higher frequency. If they arrive less frequently (longer wavelengths), we say they have a lower frequency.

In normal transmission, waves usually arrive at the same frequency they are sent. Not so when that transmission is between the surface of a collapsing star and a ship such as ours stationed at a distance. Regardless of whether radio waves, light waves or any other sort of waves are stretched by acceleration or by gravity or by both, the stretching results in the wave crests reaching us in our ship less frequently (at a lower frequency) than that at which they are being sent. What can be said for wave crests is also true for pulses of light, beeps, or anything else coming from the star. The frequency of the waves or pulses, and thus the frequency of the events we perceive by means of these, gets lower and lower. What we perceive is a slowing down of time. If the redshift is infinite, as we've said it is for waves or pulses originating at the event horizon, then, as far as we are concerned at the receiving end, time at the event horizon stops.

Returning to our specific situation on board the spacecraft. The signals from the astronaut's communicator were our indicator of the passage of time on the surface of the star. We know the clock was timing it all very carefully and sending a signal regularly every second. If the signals came further apart, *we* could only interpret that to mean that seconds down there were lasting a little longer! The second just before 4:00 (by our clocks) lasts forever.

How could this happen? It happened because the only experience we could have of what was happening on the surface of the star came from what the signals could tell us. Because their arrival was increasingly delayed, they told us that the speed at which events occurred on the surface was slowing down. Things there, by this report, were taking longer and longer to happen. Seconds were getting longer. Time was moving more and more slowly.

The increase in the distance between the signals at our receiving end was small at first – not something we could detect, though some of our ship's instruments undoubtedly could. As the gravitational pull at the star's surface and the acceleration of the star's surface away from us increased, so did the delay between the signals. Eventually, between the 3:59:58 signal and the 3:59:59 signal, the delay was great enough for us to notice without the aid of sensitive instruments. The interval between the 3:59:59 signal and the 4:00 signal is forever. From our vantage point on the ship, time at the event horizon never reaches 4:00. It has come to a complete standstill.

By the same token, the interval between the wave crests of the light in the light signals was also stretched. The light was redshifted. The result – the light we saw for the last visible signal was reddened. For the 4:00 flash, sent from the event horizon, the redshift was infinite. We saw nothing.

We know from the two contradictory statements above that the sacrificial astronaut on the surface of the star experienced no slowing down of time at all. Why not? Because the intervals between the signals were not stretched right there where the astronaut was, where they started out. We have said that because of the Doppler effect the pitch of a locomotive whistle drops as the locomotive passes and moves away from you. However, if you are riding on the locomotive, you don't hear a drop in pitch. As another analogy, imagine standing at the bottom of a very deep pit. The steepness of its sides does not slow you down until you try to climb out. The intervals between the crests of light waves, similarly, are not stretched until they start to climb away from the black hole, out of the 'pit'. The same goes for intervals between pulses or beeps. The astronaut, down in the pit, doesn't notice them slowing down any more than the engineer of the locomotive hears the drop in pitch. Nor does the astronaut perceive that during the fraction of a second it takes to cross the event horizon, our clocks on the ship show that billions upon billions of years pass by.

Is time dilation all an illusion, or perhaps just a breakdown in communication? No, it might seem to be only that, but it is more. There is a small gravitational redshift and time dilation from the bottom to the top of a skyscraper. The fact that time runs at different rates, depending upon one's distance from the source of gravity, shows up even over such a short distance. If you and I synchronized extremely accurate watches (far more accurate than we are likely to own), and then I went to the top of the skyscraper while you stayed on the ground floor, after a few hours we could meet again, compare watches and find that yours was a tiny fraction of a second behind mine. In the operation of super-accurate navigation systems based on signals from satellites, if we ignore time dilation, our calculation of position can be off by several miles.

After the star has become a black hole, there is no more redshift resulting from acceleration, but the gravitational redshift continues and so does its effect on time. When we on the spacecraft measure it by our watches, we find time nearer the black hole enormously slowed down compared with where we are. From our point of view, we see that at the event horizon it stops. But if we were to sit just outside the event horizon and keep in contact with events back at home on earth, we would see them moving forward at a pace that would amount to a blur.

On first thought, it would seem that if time stops at the event horizon (from our ship's point of view), then we ought to continue to see the star's image on the viewing screen, frozen at the event horizon. Instead, we saw the image freeze and then fade. Shouldn't we at least be seeing light that was emitted just before the event horizon formed, light whose arrival at our ship has been delayed by time dilation? Shouldn't we be seeing that for many years, possibly almost forever?

Replay in slow motion

When we play the videotape of the collapse in slow motion we notice much more detail than we observed before. (If the star was rotating before it began to collapse, this scenario becomes too complicated for us to deal with here. Therefore, for present purposes, let us agree that the star we are observing was not rotating. Also, this description takes into account only what is visible, not what we might observe with instruments capable of studying radiation outside the visible part of the spectrum.)

As the video replay begins, we see a bright disk, shrinking but not growing noticeably dimmer or redder. The rate of shrinkage

Photons from a star do not all come out radially (on pathways that look like the spokes of a wagon wheel). They emerge at many different angles, on pathways more like the spokes of a sports car wheel.

Spokes of a simple wagon wheel come out radially from the hub.

Spokes of a sports car wheel come out at many different angles from the hub.

Figure 5.2.

is slow at first, soon speeds up steadily, then stops. The much reduced disk is frozen on the screen. At its centre a dark red spot appears and grows darker until it's black. We see this spot eat its way out into the bright rim. As this happens the rim gets thinner and thinner and eventually becomes invisible.

How do we account for this scenario?

We might expect to find that all photons (particles of light) emitted by a star come straight out – radially – like spokes of a simple wagon wheel (Figure 5.2). However, a little thought should tell us that would not be the case. A star radiates light in many different directions and at many different angles from its surface. Most photons come out at angles that cause their paths to look like some of the spokes on the wheel of an expensive sports car. These latter (let us for convenience refer to them as the sports-car-wheel photons) have less chance of escaping than those coming out radially (wagon-wheel style). As the star collapses, even before the event horizon forms, spacetime curvature bends the paths of some of the sports-car-wheel photons all the way back in – something it cannot do at this stage with those photons that are coming out radially. Other sports-car-wheel photon paths are bent not quite so far as that, but far enough to be captured for a while in a circular orbit around the collapsing star. A circular orbit is not a stable orbit, and the photons in that orbit leak out gradually, some of them on paths that lead to our eyes. We see that leakage as the bright rim in the image.

In the meantime, there are photons coming out at us radially (wagon-wheel style). They are not drawn back in or detained in the circular orbit. They reach our eyes more quickly, but we see relatively few of them from any one position out in space. When we look at the edge of a wagon wheel, very few of the spokes are pointing directly at us. Soon (while sports-car-wheel photons are still leaking from the circular orbit, allowing us to continue seeing the 'rim'), there aren't any of the wagon-wheel photons left except some that are too redshifted to be seen at all – redshifted beyond the visible spectrum. No more photons in the visible spectrum coming to us means blackness, and we observe the dark dot in the centre of the disk because no more visible photons are reaching our eyes from that direction. The rim lasts longer than the centre, but the image becomes increasingly dim until it is invisible, and finally we see nothing at all.

If you are particularly intrigued by this description and want to understand it better, you can find it described in an article by Kip Thorne, of the California Institute of Technology, in *Scientific American*, November 1967. Keep in mind that the 'rim' we see is not the photons in orbit, but those that are leaking from the circular orbit on paths that reach our eyes. We are not able to see any photons whose paths do not end in our eyes. If a photon were able to orbit forever, it would *never* be seen by a distant observer.

It was not time dilation, then, that caused the image to freeze on our screen. It's true that time dilation slowed down our picture of the collapse, but, until the light was redshifted beyond our visible spectrum, not greatly enough to freeze the image. And – yes – there is probably delayed light still reaching our ship, light that escaped just before the event horizon formed, and some of this light may take nearly forever to reach us, but it is redshifted beyond the visible spectrum.

The fate of the astronaut

We have been deliberately ignoring, for the sake of our own understanding, the fact that the unfortunate astronaut on the surface of the star has had little opportunity to appreciate the experience. Tidal effects surely strung the volunteer out like spaghetti and tore him or her apart, probably proving fatal even before the event horizon formed. In some state or other, perhaps reduced to fundamental particles, though even they probably cannot hold together in this situation, the astronaut reaches the singularity, the point of near infinite density, where spacetime curvature and the bending of paths of light are also near infinite – the end of space and time as we know them.

We can learn more by turning our attention to an astronaut falling into a much larger black hole. Holes sufficiently large for the following description may exist at the centres of quasars and active galaxies. They might not exist in other locations where it would be hypothetically possible for our ship to visit and a volunteer to jump in. However, let us for the moment say that they do.

Even above the event horizon of such a black hole there would be some impressive and daunting visual effects. Beneath the astronaut, the view would be dank and forbidding, a pool of blackness. As the astronaut fell, that blackness would spread and then appear to creep up the edges of the sky all around and eventually above the astronaut, until light was visible only through a tiny, bright, circular hole directly overhead. All the galaxies of the universe, even those that are on the other side of the black hole from the astronaut, would appear up there, many of them double and triple imaged, in that small bright area. It would be like standing at the bottom of a deep, narrow, dark pipe, seeing a tiny, distant disk of light at the end far above you. In Chapter 6 we'll talk more about such effects and why they would occur.

There would be no way for the astronaut to recognize the event horizon – no line or surface or abrupt change in the way things looked or felt. There would also be no physical danger or discomfort here or perhaps for a while after that (by the astronaut's watch) – no pasta-making by tidal gravity in such a large black hole for some distance beneath the horizon. The astronaut would be too far from the singularity to experience its tidal gravity in a significant way. The gravitational pull here would not differ enough over the length of a human body. Meanwhile, signals would still be able to reach the astronaut from outside the hole, because, of course, light and other radiation can *enter* a black hole. The event horizon is only a one-way barrier.

You may be wondering whether, with conditions such as these at the event horizon, it might be possible to escape. Suppose someone lowered you on a strong tether past the event horizon. Could they pull you back out? A ten-year-old asked me that question, which his friend answered before I could: 'No, because I would cut the rope!' The answer is indeed 'No,' not because someone would cut the rope but because your journey out would still have to take place at a speed greater than the speed of light, and one principle of relativity is that nothing can move faster than the speed of light. You would have the same insurmountable problem if you tried to increase your chances of escape by firing a rocket. Bungie jumping at the event horizon is also not recommended.

Getting back to the astronaut, he or she would continue to fall, moving more and more rapidly, and the tidal effect would grow stronger. At first the stretch from head to foot and squeezing from the sides would be only mildly disagreeable, but all too soon they would grow more severe, and when muscles, bones and tissue could no longer withstand the stretch and squeeze, the astronaut would perish. Even the atoms that made up the astronaut's body and the elementary particles that made up the atoms would finally be torn apart.

We said earlier that tidal effects could stretch an astronaut out 'like spaghetti'. The final result could actually be rather different. Near the singularity, it's likely that tidal gravity oscillates chaotically. If so, it would pull and squeeze different parts of the astronaut's body first one way and then another like a mechanical taffy-pulling machine gone berserk. Indeed, as all of this activity grew faster and stronger, the astronaut would be whipped into something more like zabaglioni than spaghetti. Why happenings around and inside black holes always seem to take a gastronomical turn, while the situation is anything but appetizing, is a question not yet addressed by science!

At the singularity, does tidal gravity grow infinitely strong, as relativity theory predicts, and the chaotic oscillations infinitely rapid? The laws of quantum physics (laws which govern the level of the very small – molecules, atoms, and elementary particles) do not allow this to happen, and we're not sure what happens instead. But whether or not tidal gravity and the rapidity of the chaotic oscillations are actually infinite will not change the fate of the astronaut.

Even if survival were possible at the singularity, there is almost certainly no way to pass through it or get beyond it. In fact, 'through it' and 'beyond it' probably make no sense at all there. This is a region we believe is ruled by the laws of quantum gravity, a merging of the laws of quantum physics and general relativity. We don't yet know the laws of quantum gravity. Those, we like to think, will be early twenty-first century physics. Our best understanding at present is that, here at the singularity, time ceases to exist as the chronological time we know (where past, present, future, before and after are meaningful words), and all becomes a random, sudsy froth. Wheeler has named it 'quantum foam'.

It could be that in older black holes, especially those that haven't eaten a star or an astronaut lately, things may have settled down and become calmer. It is possible that in this somewhat gentler environment the astronaut might survive right up to the edge of the singularity. This would be interesting, if not exactly fun. Canadian physicist Werner Israel suggests that just before the astronaut reached the core of the black hole there would be 'a blizzard of messages from the outside world'. Meanwhile, a 'fine drizzle of gravitational waves would hit like a pneumatic drill.' The astronaut would probably be pulverized by the radiation and reduced to nothing but a flash of gamma rays.

Not everyone has completely given up hope of survival in such circumstances, or relinquished the feeling that there just must be a way to get beyond the singularity. Israeli physicist Amos Ori (now at Cal Tech) believes there is a small chance the wall of infinite density surrounding the core might be thin, like an eggshell, which would allow the astronaut to be swept rapidly through a wormhole to another universe or to another place or time in our own universe. 'You see the crazy ideas we're playing with', Werner Israel comments. We can hope our own intrepid volunteer was not relying too heavily on any such possibility, not only because the hole we are visiting is much too small to allow survival even at the event horizon, but also for other reasons we shall consider when we return a little later to the subject of wormholes.

6

Contemplating an enormous nothing

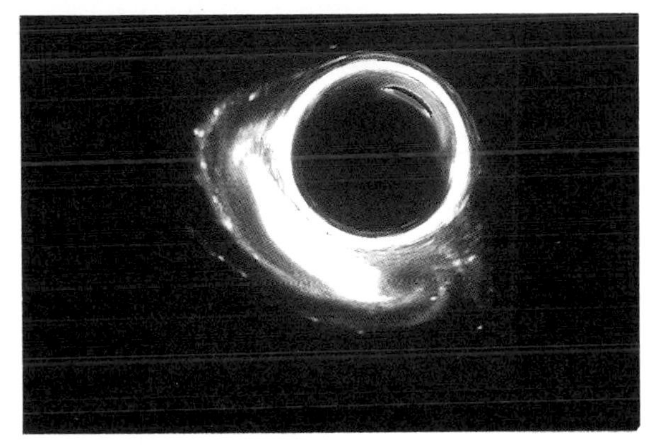

Computer simulation of (above) a galaxy as it would normally appear and (below) the same galaxy as it might appear if a black hole came between it and us. The paths of light from stars in the galaxy are bent around all sides of the black hole at once. They reach our eyes as a bright ring, known as an 'Einstein ring'. (Courtesy of Emilio E. Falco, Smithsonian Astrophysical Observatory.)

A huge great enormous thing, like – like nothing. A huge big – well, like a – I don't know – like an enormous big nothing . . .

Piglet describes the Heffalump,
in *Winnie the Pooh* by A.A. Milne

The drama of the star's collapse is over. We've reviewed the video in slow motion and speculated about the fate of the volunteer. The time has come to turn our attention to the new black hole. This may seem a little problematical. We can't see anything, can we?

What can we see?

Nothing can come out of the black hole and we can't actually see the hole, but that doesn't mean there isn't evidence that it's there. Atoms of gas that form the thin interstellar medium – an extremely sparse population of atoms out there in space – drift toward the black hole from all directions, barely creeping along out where the hole's gravitational pull is weak, moving more rapidly nearer the hole, rushing at almost light speed very close to it. We cannot see these atoms, but with instruments aboard our ship that study different types of electromagnetic radiation we can study the radiation the gas emits. Far out from the hole where the gas is still cool it emits radio waves. Closer in, where the atoms in the gas are moving faster, they collide and heat up. The heat makes them vibrate more rapidly and they emit waves of shorter wavelength – wavelengths in the visible part of the spectrum. The display here is quite colourful. Still closer to the hole, the atoms move faster yet. Collisions heat them until they vibrate even more rapidly and emit X-rays. Still closer, they emit gamma rays. From our ship we can follow this progression with our various instruments, making allowances for the redshift, until . . . darkness – where the radiation, though still being emitted at very high energy, becomes too redshifted (from our point of view) for any of our instruments. For radiation emitted at the event horizon the redshift is infinite. After the atoms have entered the black hole, they continue to emit radiation, but no radiation can escape from the hole. The nearest we get to 'seeing' a black hole directly is to see it as an absence of anything to be seen.

Depending on our ship's distance from the black hole, even the part of this display in the visible spectrum may not be easy to make out. We might in fact think we are looking at a rather ordinary starry sky. However, we mustn't be so quick to assume that this starry sky is 'ordinary'. The reasons we may be fooled into thinking so are twofold. First, if the black hole is small enough relative to our distance from it, it creates an area of darkness not noticeably larger than other areas of darkness nearby. Second, a black hole doesn't merely blot out a portion of sky but instead employs a cloaking device: it bends the paths of light from stars behind it in a way that distorts our picture of the background sky and gives us a false impression about where the stars are.

William Unruh of the University of British Columbia in Canada has created a computer-generated film of this effect. Unruh's view is from a spacecraft in orbit around a 1-billion-solar-mass black hole, considerably larger than the one we are contemplating. In Figure 6.1 we have two frames from Unruh's film and a Key to help us locate the stars. Frame A shows the area of sky containing the constellation Orion the way the astronauts would see it on their viewing screen if there were no black hole present. Frame B shows the same area of sky on the viewing screen with the black hole between the spacecraft and the constellation.

In Frame A, we can spot the Orion belt stars near the centre. Among the stars below that and slightly to the right, the brightest star is Rigel. Among the stars above the belt stars and slightly to the left, the brightest is Betelgeuse. These and other familiar stars are labelled in the Key. Turning our attention to Frame B and its Key, we see a different picture. There are primary and secondary images of the belt stars and others, and distortion of their positions relative to one another.

Figure 6.2. helps clear up the mystery of how such distortion happens. Diagram (*a*) shows the paths of five particles coming from the right of the drawing. The paths bend as they encounter the

Frame A

Frame B

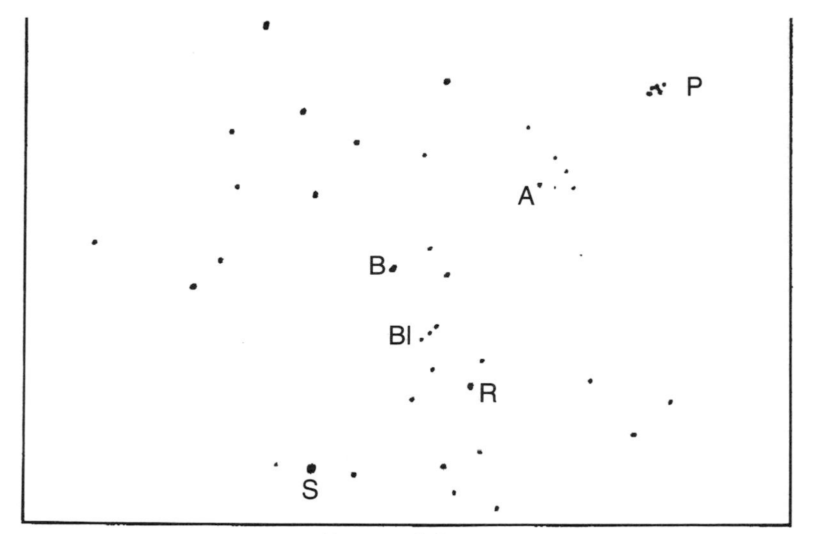

Frame A key

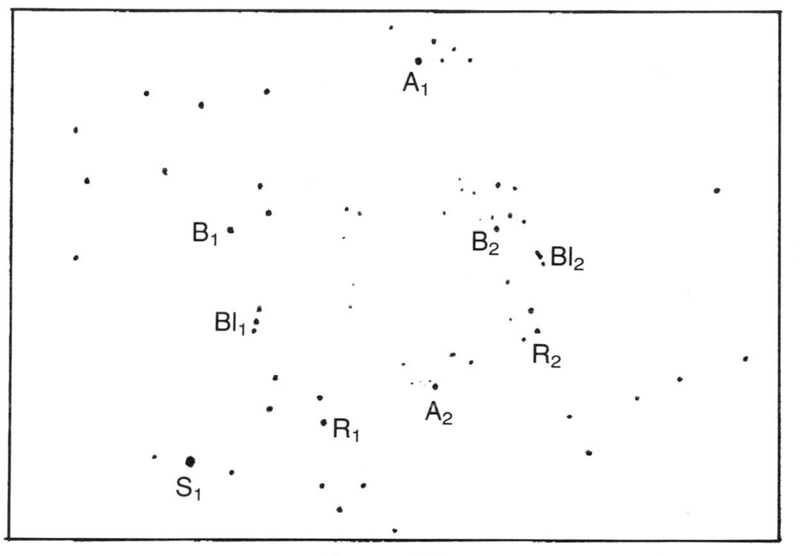

Frame B key

Figure 6.1. Two frames from William Unruh's computer-generated film showing (A) a view of the sky around the constellation Orion, and (B) the same area of sky when a black hole has come between the viewer and the stars. In Frame A we see the belt stars and other familiar stars (labelled in the Key). In Frame B there are primary and secondary images of the belt stars and other stars (again, consult the Key). (Courtesy of William Unruh.)

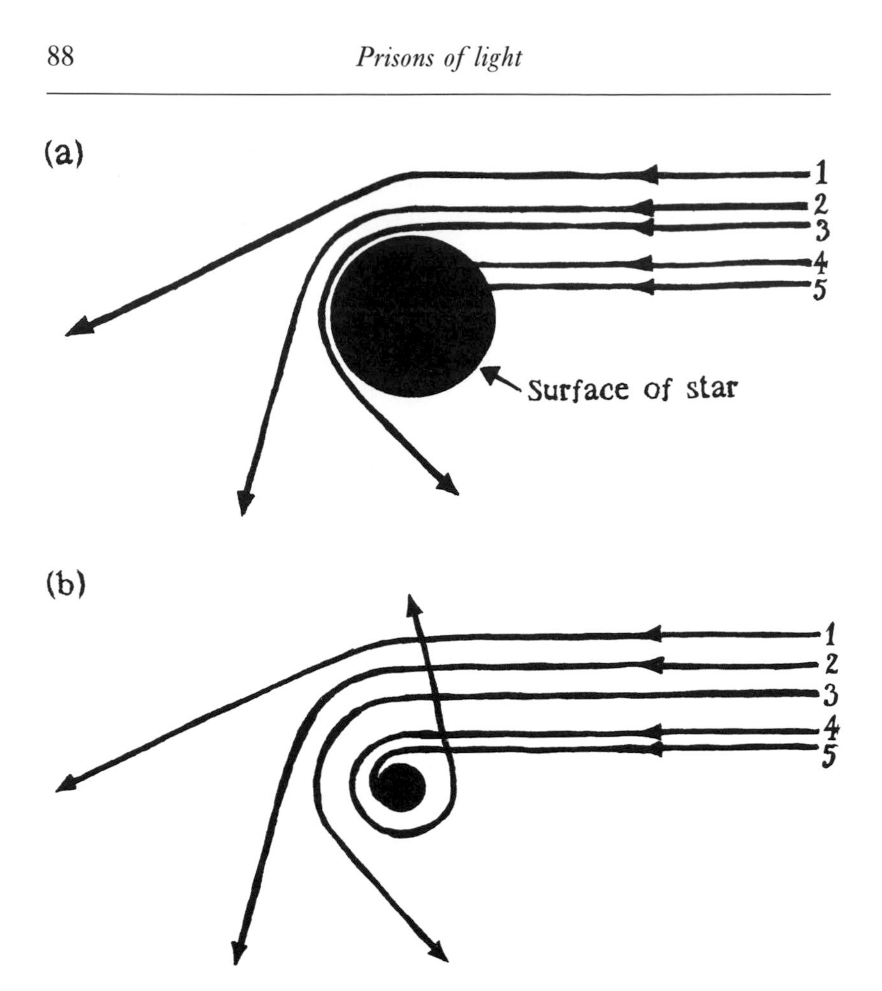

Figure 6.2. (*a*) Five particles move toward a star. The paths of particles 1, 2 and 3 are bent as they pass the star. The closer to the star, the greater the bending. Particles 4 and 5 hit the star. (*b*) The same particles move toward the star after it has become a black hole. The paths of particles 1, 2 and 3 are bent exactly as before, because the spacetime curvature at these distances from the centre of the star are the same as at these distances from the centre of a black hole of the same mass. (Recall the squeezed earth.) Particle 4 circles the hole, perhaps many times, and then escapes. Particle 5 is drawn into the hole. (Courtesy of Clifford Will.)

warp of spacetime around a massive object, a star. As we would expect, remembering the bowling ball on the trampoline, the closer to the star the greater the effect. The paths of particles 4 and 5 hit the star. In (*b*) the star has become a black hole, but it still has the same amount of mass it had when it was a star. Remembering the 'shrinking earth' story, we will not be surprised to find that the bends in the paths of 1, 2, and 3 are the same as they were in (*a*), before the star collapsed. The paths of 4 and 5 also have not changed, but, because the star has shrunk, they no longer hit it. Instead, they enter areas of extreme spacetime curvature. Particle 4 circles the black hole and escapes. Particle 5 falls into the black hole. Looking at this diagram, it is not too difficult to accept that the paths of light from a distant star could be warped by the presence of a black hole in the foreground so as to produce double images and give us false impressions of the star's position and distance. Nor will we be surprised to learn in later chapters that such double images and distortion of the background – such 'lensing effects' – are clues when astronomers look for otherwise undetectable accumulations of matter in the universe.

On the viewing screen in our imaginary ship we may see another distortion that Einstein predicted. If a star or a galaxy is precisely centred (relative to us, the observers) behind the black hole, the paths of some of its light may be bent around all sides of the black hole at once. The starlight reaches us as a bright ring. The simulation of this effect in the frontispiece of this chapter was created by Emilio Falco of the Smithsonian Astrophysical Observatory.

We'll move in closer now to the black hole and try some simple experiments. We'll turn on the ship's most powerful outside spotlights and direct them at the centre of the area of sky where the black hole is but can't been seen. Will that light reflect off anything inside the black hole so that we'll see the reflected light? Predictably, we see nothing. Shall we try a laser beam? Still nothing. Perhaps there is nothing in the black hole to reflect the light. After all, the matter that once constituted the star has now been reduced to a singularity. The area between the event horizon and the singularity is probably empty. If we chuck something in, something that will reflect light – a chrome garbage bin from the ship's kitchen, perhaps, full of garbage so as not to be wasteful – will we see the light reflected off that? We eject the garbage bin and watch it veer off toward the black hole. After a short burst of radiation it disappears. There is no reflection. Or perhaps there is, but none that can reach our eyes. Even ignoring the effects redshift and time dilation have on our view of things, we know the photons from the spotlights and laser, reflected off the bin inside the black hole, can't move back in our direction. They are drawn to the singularity at the centre.

What can we measure?

Suppose we had stumbled upon this black hole after it had become a black hole and we had never known it as a star. What could we learn about the star that collapsed to form it? Could we determine, for instance, whether the star was made of matter or antimatter? Whether it was spherical or irregular in shape? Whether other items had fallen in as it collapsed or later? Whether an astronaut's body was part of the debris mercilessly compressed to a singularity? No. Black hole theory tells us there are only three secrets a black hole divulges: its mass, its angular momentum, and its electrical charge. All other information has been reduced to those three. John Wheeler (Figure 6.3), who invented the name 'black hole', also coined the phrase 'Black holes have no hair' to reflect this dearth of information. As he puts it, 'Not one hair remains to betray its past.' Figure 6.4 is Wheeler's whimsical drawing of an assortment of items entering a black hole, all to leave it possessing at the end – as seen from outside the event horizon – only mass, angular momentum, and charge.

We can measure these three attributes indirectly by studying their effects on spacetime around the black hole. The ability to search for black holes in the real universe and to know when we have found one depends in large part on our ability to understand and interpret these effects.

Mass

If the sun were invisible, it would nevertheless be possible to calculate its mass from the orbits of the planets. Likewise, we can

Figure 6.3. John A. Wheeler, one of the leading gravitational physicists of the twentieth century and mentor to generations of physicists, lectures to his students at Princeton University. He coined the term 'black hole'. (Princeton University/Robert P. Matthews.)

attempt to calculate the mass of the black hole by studying orbiting objects such as other stars, planets, or even our own spaceship. In the next chapter we will learn in greater detail how this line of thinking helps identify black holes in 'binary systems', where we suspect a visible star and a black hole are locked in a partnership and are orbiting around their common centre of mass.

Another way to find the mass of our black hole would be to study the lensing effects mentioned above – to what extent and in what manner light from distant stars is bent by the presence of the black hole.

Having worked out the mass of this particular black hole, we have the information we need to calculate its circumference. To do this we multiply its solar mass by 18.5 kilometres. A 100-solar-mass black hole has a circumference of 1850 km.

Does this mean we now know the mass the star had before it collapsed to form a black hole? We must be cautious here, remembering that the mass of the black hole is not necessarily the same

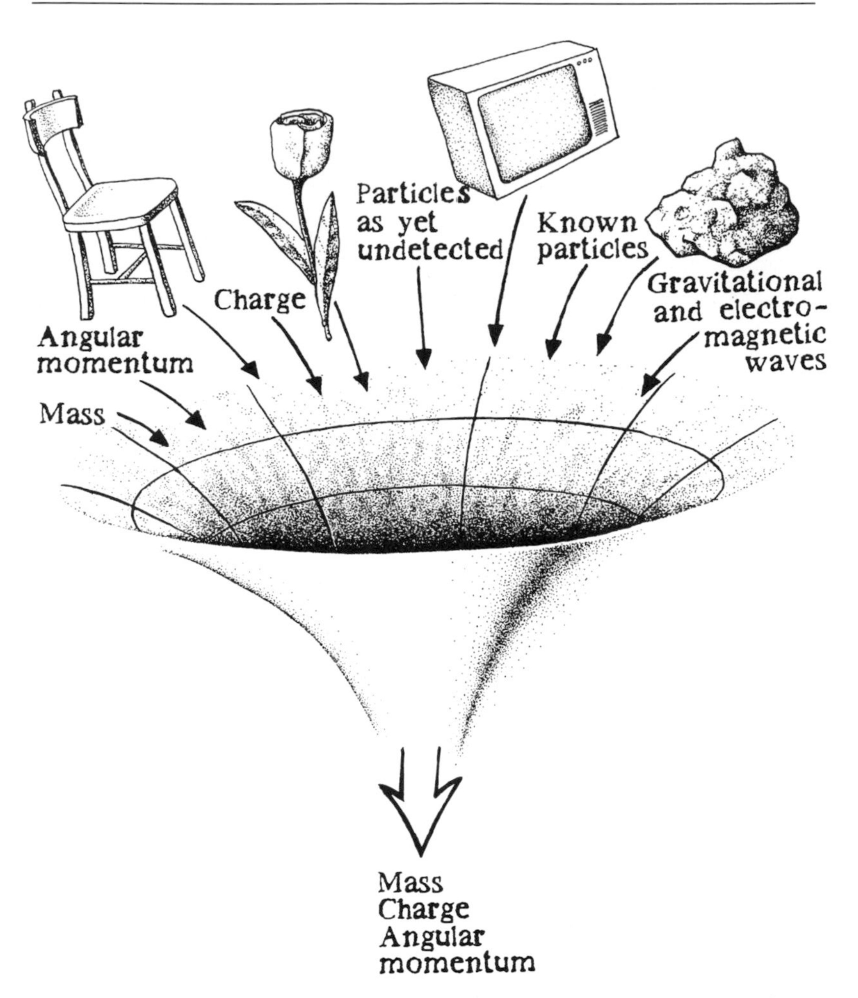

Figure 6.4. 'Wheeler's funnel' shows an odd assortment of objects falling into a black hole. They leave it (as seen from the outside) possessing only mass, angular momentum, and electric charge. (Courtesy of John A. Wheeler.)

as the mass of that star. A portion of the star's mass may have been carried off late in the star's life in explosions or in the turbulence of the collapse. Ageing, dying stars sometimes eject enormous amounts of mass. We may see some of the debris still orbiting the black hole. All we know is that whatever was left of this star when the event horizon formed is now inside the black hole, compressed to near infinite density at the singularity.

Angular momentum

We have calculated the mass of our black hole, and we know that the mass is concentrated at its centre, at the singularity. Our next task is to figure out the black hole's rate of rotation and its angular momentum. At the beginning of this chapter we talked about atoms of interstellar gas falling into the black hole. If the hole is rotating, these atoms will be swept into a swirl around the black hole as they near the event horizon. We can attempt to find out the hole's rate of rotation by studying the way they swirl, or by sending out test objects of our own and noting where they also begin to be swept around irresistibly with the hole. In that way we can figure out where the ergosphere begins (see Chapter 4 and Figure 4.5). Knowing the dimensions of the ergosphere allows us to calculate the black hole's rate of rotation, remembering that the greater the rate of rotation, the larger the ergosphere. If there is no ergosphere, the black hole is not rotating at all. The interstellar gas atoms and our test particles will not be swept around. They will plunge straight in. We ought now to be able to draw a picture of our black hole. If we have found that it's rotating, our picture will show it bulging at the equator, and its shape will be elliptical. The greater the rate of rotation, the greater the bulge. If the black hole isn't rotating, it will be spherical. Though black holes come in an enormous range of sizes, it seems these are the only two shapes available – spherical and elliptical (Figure 6.5).

Knowing the angular momentum of the black hole, can we calculate the angular momentum of the star that collapsed to form it? Not with any certainty. Although angular momentum is conserved and could not disappear as the star collapsed, during the collapse some of it may have been carried off or transferred to other objects.

Electrical charge

An atom has equal amounts of negative and positive charge. A star consists of atoms, so we expect it also to have equal amounts of

Three types of black holes, named after the mathematicians who first worked out their theoretical existence.

Schwarzschild black hole: spherical in shape, non-rotating, no electrical charge. (Karl Schwarzschild, German astrophysicist)

Reissner-Nordström black hole: spherical in shape, non-rotating, has an electrical charge. (Physicists Hans Reissner and Gunnar Nordström, German and Dutch respectively)

Kerr black hole: not spherical in shape, rotating, with or without electrical charge. (Roy Kerr, mathematician from New Zealand)

Figure 6.5.

negative and positive charge. Most theorists conclude that, as a star collapses, these probably cancel one another out, leaving the black hole with a net charge of near zero. If a black hole did contain much electrical charge, it would almost immediately pull opposite charges into itself from the interstellar gas and thus neutralize its charge, ending up with a net charge of near zero.

Can anything escape?

So far in this book, although we've spoken a little about the activities of elementary particles and atoms, our discussion has not strayed far into quantum theory, the theory having to do with what happens on the level of the 'very small' – molecules, atoms, and elementary particles. We are now about to venture into this more exotic theoretical territory. For physicists studying black holes, this change of focus was triggered by a question about the entropy of a black hole, and that is the route we shall take. First, what is entropy?

Measuring the amount of entropy in the universe – or in any system – means measuring the amount of disorder or randomness. A law of physics known as the Second Law of Thermodynamics has it that entropy (disorder) in any 'closed system' can never decrease, only increase. Things can only get more disorderly. This law does not always hold, but entropy in the universe as a whole is generally accepted to be inexorably on the rise.

Common sense may scoff at such a conclusion. If we place marbles of two colours in a box, separating the reds from the greens with a partition, and if we then remove the partition and give the box a shake, there is only the smallest of probabilities that at any future time the marbles will again sort themselves out by colour as they began. If we saw them do it, we wouldn't believe our eyes. However, if we reach into the box and re-sort the marbles, doesn't that defeat the Second Law of Thermodynamics? No. If we reach into the box and interfere, box and marbles are no longer a closed system but part of a larger system that includes us. Any physical and mental effort we expend to put a part of the universe in order, marbles in boxes, our homes or gardens, converts energy into a less useful form, and this increases the overall entropy of the universe. Nor can we foil the Second Law by lying around doing absolutely nothing. Merely staying alive converts energy.

One way to see this problem is to recognize that in any system the start-up conditions that would allow things to become more ordered are vastly rarer than the start-up conditions that would allow them to become less ordered. The marbles in the box would all have to be rolling at just the right speeds and in just the right directions to get back to their sorted positions on the two sides of the box. It could happen, but it isn't at all likely given the much greater number of other speeds and directions that *wouldn't* get them there.

When entropy (disorder) increases, things get less predictable, and for this reason an increase in entropy represents a loss of information. One definition of entropy is 'lack of knowledge about the precise state of a system'.

But suppose we toss the box of mixed-up marbles (or anything else that has entropy) into a black hole. Goodbye to that bit of entropy, we might say. The total amount of entropy in the universe is less than it was before. We have succeeded in defying the Second Law of Thermodynamics! Or have we?

Tea with John Wheeler

On an afternoon in 1970, Wheeler was having tea in his office at Princeton with a graduate student, Jacob Bekenstein. Wheeler lamented that mixing hot and cold tea was a crime whose consequences 'echo down to the end of time', because the transfer of heat increases the disorder of the universe, the information loss, the entropy. The increase is an irreversible process. Wheeler's crime can never be undone. 'But', Wheeler continued, 'if a black hole swims by and I drop the teacups into it, I conceal from all the world the evidence of my crime.' There is no increase in entropy.

After giving the matter some thought, Bekenstein came back to Wheeler with the bad news that his scheme wouldn't work. Dumping the tea into the black hole would merely result in an increase in the entropy of the black hole.

This sounds like a whimsical anecdote and an answer any one of us might have come up with – since to most of our minds a black hole would seem a veritable repository of lost information! But Bekenstein's suggestion was a radical one, because black holes were not supposed to *have* entropy. Bekenstein insisted they must. In his way of thinking, the area of the event horizon of a black hole isn't only *like* entropy, in that neither can ever decrease. It *is* entropy. Measuring the event horizon is measuring the entropy of the black hole.

What made this idea so difficult to swallow within the physics community was that if something has entropy, it also has a temperature. It is not stone cold. If something has a temperature, it is radiating energy; and if it is radiating energy, we can no longer say that nothing is coming out of it. Nothing was supposed to come out of black holes!

Complicating matters was the fact that this same year Stephen Hawking discovered the Second Law of Black Hole Dynamics, a discovery so simple (by Hawking's standards) that with hindsight it seemed to him that anyone could have thought of it, but one which excited Hawking so much that he lay awake for most of the night pondering it. The discovery had to do with those photons that hover at the event horizon of a black hole, neither escaping nor falling in (see Chapter 3). As Hawking describes it in *A Brief*

History of Time, he realized that the paths of light rays hovering there cannot be paths of light rays that approach one another. If they were, they would collide with one another and fall into the black hole, not hover as paths of light at the event horizon must do. If the area of the event horizon got smaller, paths of light rays in the event horizon *would* approach one another, and when they collided they would fall into the black hole. To Hawking's way of thinking, this would mean they could not have been on the boundary of the black hole. If the rays of light that form the event horizon can never approach one another, but must always be moving parallel to one another or away from one another, the area of the event horizon may stay the same or increase but can never decrease.

Didn't we reach that same conclusion by a different route in Chapter 3? Almost. We said that where the event horizon is (the circumference at which it is) depends upon the mass of the black hole. The mass of a black hole increases if anything falls in, but how could it possibly decrease if nothing can get back out – if the black hole can't lose any mass? Can the mass ever grow less? It would seem not, and, if not, doesn't it follow that the area of the event horizon can't decrease either? We left those questions hanging in Chapter 3. We must return to them now and come to grips with some complications.

First – in one sense quite apart from whether a black hole has entropy and a temperature or whether anything can escape from inside – a black hole *can* lose mass. You may recall from Chapter 2 that there is an 'equivalence' of mass and energy embodied in Einstein's equation $E = mc^2$ which allows us to treat mass and energy as two forms of the same thing. In that equation the E stands for energy; the m for mass; the c for the speed of light (Figure 6.6). When the energy E (on one side of the equals sign) changes, something on the other side of the equals sign must change too. The speed of light c cannot change. The mass m is the only thing there that can increase or decrease. Hence, any loss of energy is a loss of mass. Now if a black hole is rotating, we can define the mass/energy of the hole to include not only the mass of what is in it but also the energy of the hole's spin. So if the spin slows down – and we will see instances where that may

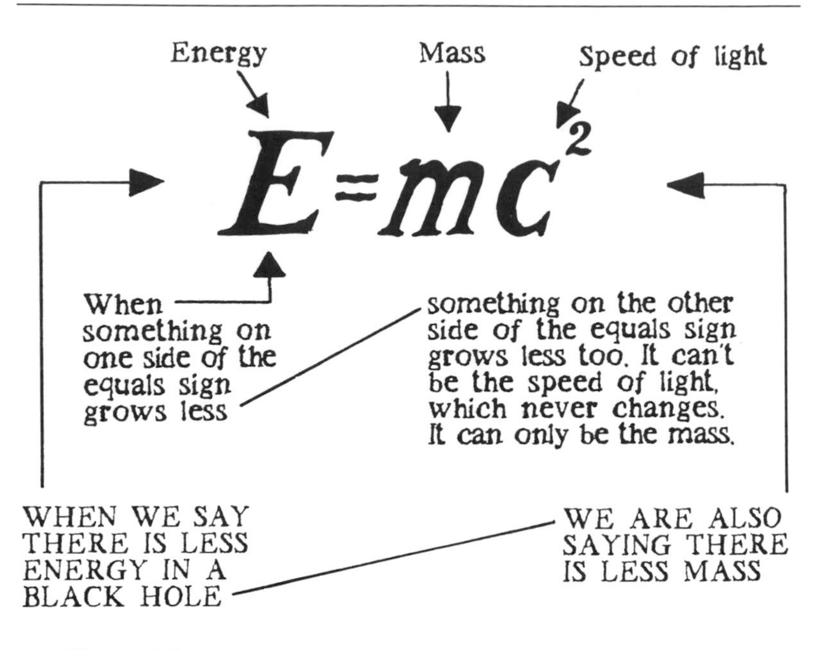

Figure 6.6.

happen – the hole's total mass/energy grows less, though not less than it would be if the hole were not spinning at all.

Continuing to think of the hole's total mass/energy, not just the mass of what's 'inside it', consider a case where two black holes collide – one of the situations to which Hawking's Second Law of Black Hole Dynamics applies. Some of their mass/energy gets converted to gravitational wave energy and is radiated away. The new black hole resulting from the collision has less mass/energy than the two previous black holes combined.

Given the discussion in Chapter 3, we might think that means the area of the new event horizon would surely be smaller than the area of the two previous black holes combined. Not so, says Hawking's Second Law of Black Hole Dynamics, which insists that even though there is less mass/energy, the *area of the event horizon* cannot decrease. If two or more black holes collide to form one black hole, regardless of the mass/energy radiated away as gravitational wave energy, the area of the new event horizon is as large as or larger than the previous event horizons added together, never smaller than that. With the Second Law of Black Hole Dynamics Hawking showed that a black hole cannot get smaller, nor can it be destroyed or split in two, no matter how hard it gets zapped. That was Hawking's conclusion in 1970. However, the story doesn't end there.

The resemblance between this Second Law of Black Hole Dynamics and the Second Law of Thermodynamics did not go unnoticed, but it wasn't until Bekenstein began thinking about Wheeler's teacups that anyone insisted that entropy and the event horizon are not merely similar in that neither can ever decrease. The area of the event horizon of a black hole *is* the entropy of the black hole. A black hole does have entropy, no matter how little theoretical physicists may like it! Indeed, it *is* a repository of lost information.

Hawking was certain that Bekenstein was mistaken, and he was irritated by what he saw as Bekenstein's flagrant misuse of the Second Law of Black Hole Dynamics. Hawking and colleagues Brandon Carter and James Bardeen set out to disprove the notion that black holes have entropy and temperature and that anything can escape.

Hawking had been looking at black holes with the eyes of cosmology, the study of the very large – stars, planets, galaxies, and the universe as a whole. In 1973, he shifted ground to look at what happens at the surface of a black hole through the eyes of quantum mechanics – the study of the very small, of molecules, atoms and elementary particles.

When Hawking was visiting Moscow that year, two Soviet scientists, Yakov Borisovich Zel'dovich and Alexander Starobinsky, convinced him that a principle of quantum mechanics known as the uncertainty principle (which we'll talk more about in a moment) means that rotating black holes create and emit particles. Hawking was not completely satisfied with their calculations, however, and set about trying to devise a better mathematical treatment, expecting to find nothing more than that rotating black holes did produce the radiation the Russians predicted.

What he discovered was far more disturbing to him and to the rest of the physics community when they heard about it. 'I found, to my surprise and annoyance', says Hawking, 'that even nonrotating black holes should apparently create and emit particles at a steady rate.' In other words, even black holes that weren't rotating could lose mass/energy. Hawking was indeed irritated, and eager than Bekenstein not find out about this discovery and employ it as an argument supporting his ideas about entropy and event horizons. Hawking was sure something was amiss with his own calculations, and he spent many more hours attempting to discover his error. Eventually he had to admit that his calculations could not be far off the mark. The clincher was that the spectrum of the emitted particles was what one would expect from any hot body.

Bekenstein was right about Wheeler's cups of tea. Throwing them into a black hole would not make the universe more orderly. As matter carrying entropy enters a black hole, the area of the event horizon increases; the entropy of the black hole increases. The total entropy of the universe both inside and outside black holes doesn't decrease.

This leave us with the problem of explaining how a black hole can possibly have a temperature and emit particles if nothing can escape past the event horizon. *Can* something escape? Hawking discovered the answer in quantum mechanics. In order to follow his thinking, we'll digress to pick up a little background in that area.

Rules in the quantum playground

In the solar system, the planets move round the sun in elliptical orbits, orbits that are not so unfailingly predictable as earlier scientists thought but predictable enough to leave us not entirely at sea. Information about where, for example, the planet Saturn is right now in relation to the sun and the other planets, and what the speed and direction of Saturn's movement are, allows us to make fairly reliable statements about where Saturn was at some specified time in the past, where it will be going next, and what path it has taken and will take to get from one position to another. Our spacecraft's pilots could plot a course and be reasonably confident that at certain spacetime coordinates we would intercept the planet Saturn.

Early in this century, scientists thought atoms would turn out to be something like miniature solar systems, with electrons orbiting the nucleus as neatly as planets orbit the sun. This was the model of an atom that we saw on the left in Figure 1.2 in Chapter 1. It was not until the 1920s that physicists found that this is not a good picture of an atom. Though no mental image really suffices, we do better to visualize the electrons blurred in a cloud around the nucleus (the diagram on the right in Figure 1.2). Why this fuzziness?

Unlike a planet in a solar system, an electron or other elementary particle never has a definite *position* AND a definite *momentum* at the same time. What that means is that we may measure very precisely the whereabouts of a particle, but we cannot simultaneously measure very precisely how it is moving. Or we may choose instead to measure precisely how it is moving (momentum), in which case we cannot simultaneously measure precisely its whereabouts (position). A helpful mental image is to think of the two measurements – position (where the particle is) and momentum (how it is moving) – riding at opposite ends of a see-saw. As the precision of one measurement increases, the precision of the other inevitably decreases, and vice versa. It is impossible to break the see-saw board in the middle, or remove the wedge that supports the middle, in order to pin both measurements down simultaneously. One or the other or both are always 'up in the air' – fuzzy (Figure 6.7).

What we have been describing is the Heisenberg uncertainty principle of quantum physics. The majority of scientists have long since given up hope of discovering a way around it, though some are still trying. With this frustrating realization, it seems that science outdoes science fiction, for we must now imagine a particle being in many places at once or following many paths at once, which certainly runs counter to our normal experience of nature. We have to stretch our minds to think of alternative positions coexisting in a way they cannot do in our everyday world, where a billiard ball or a planet is never in several places at once. We also can't help noticing that 'reality' on the quantum level of the universe seems to depend upon which measurement we choose to make. What, then, is 'real' when we aren't taking any measurements? Anyone who is uncomfortable with this sort of thinking

On the quantum level of the universe (the level of the very small – molecules, atoms and elementary particles), with some pairs of measurements, if we pin down one measurement very precisely, the other is 'up in the air' – fuzzy, imprecise. We can choose which to measure precisely, but we can't pin down both precisely at the same time.

How it is moving.

Momentum

Where it is.

Position

At first physicists thought this problem was only due to their inability to do the measurements skillfully enough, but later they came to the conclusion that when we measure momentum precisely, the particle does not actually have a precise position, and vice versa. In this sense, our choice of which measurement to take determines what reality is on the quantum level.

What we are dealing with here is the Heisenberg uncertainty principle of quantum mechanics. It applies to other pairs of measurements as well, such as the value of a field and its rate of change over time.

Figure 6.7. Rules for the quantum playground.

should not feel alone. The uncertainty principle profoundly disturbed the scientific community when it was first discovered and in the years following that discovery. Even now, when we have long since come to treat the uncertainty principle as one of the givens of our universe, and when quantum mechanics underlies much of our most familiar and dependable technology, no one claims to understand fully this uncertainty or how it translates into the common sense world of our own level. The philosophical debates the uncertainty principle sparked, though they do not now rage as they did in the decades following its discovery, have not ended.

Hawking's radiation

It is not so surprising, then, that it is the uncertainty principle that provides an opportunity for escape from a black hole. While the uncertainty principle precludes our ever knowing precisely both the position and momentum of a particle at the same time, it also forbids other simultaneous precise measurements. The ones that concern us here are the value of a field (for instance, a gravitational field or an electromagnetic field) and the rate at which the field is changing over time. Getting on with this book doesn't require understanding exactly what those terms mean. Suffice it to say simply that the more precisely we know the value of a field, the less precisely we know the rate of change, and vice versa. Another see-saw. It follows, then, that we will never find them both measuring 'zero' at once, because 'zero' is a very precise measurement indeed.

The upshot is that although we may think of space as a vacuum, it is no such thing, because empty space cannot exist unless all fields and their rates of change over time measure precisely zero. No zero, no empty space. What do we have instead of a true vacuum? The uncertainty principle requires that we have a minimum amount of uncertainty, a bit of fuzziness, as to what the value of a field is in 'empty' space.

This uncertainty takes the form of 'quantum fluctuations', which happen everywhere in space, at all times. During these quantum fluctuations, pairs of particles appear – pairs of photons or pairs of gravitons, let us say. The particles separate and then come together again and annihilate one another after an instance of existence lasting an unimaginably small fraction of a second. Because this all happens so rapidly, these particles never come into evidence at any everyday scale of observation. They are *virtual* particles, not *real* particles, and we know they exist only because we can measure their effects on other particles. Some of the pairs are pairs of matter particles, electrons and positrons ('anti-electrons') or protons and anti-protons, for example. In these cases, one of the pair is an antiparticle. 'Antimatter' is not science fiction.

Theory has it that the production of particle pairs will be greater where the curvature of spacetime is most severe and changing most rapidly. We can expect to find a great many particle pairs at the event horizon of a black hole. What's more, under conditions near the horizon, tidal effects of gravity can pull the pair apart with enormous force. That feeds enough energy into them to change the virtual particles into real particles. As real particles, they don't have to meet again and annihilate one another.

Here, then, is one way of thinking about the 'Hawking Process'. Two newly created particles at the event horizon exchange energy, one ending up with negative energy and the other with positive energy. The negative-energy particle falls below the event horizon to the singularity. The positive-energy particle may fall into the black hole too, but that need not happen. Because it is a real particle now, not a virtual particle, it doesn't have to rejoin its partner and annihilate. It may instead escape to a distance outside the black hole, making off with the borrowed energy – a debt it will never repay. To an observer at a distance the particle appears to come out of the black hole. In fact, it comes from just at the border. All this turns out to be the black hole's loss, for the black hole must in a sense carry the debt of energy brought in by the negative-energy member of the pair. Negative, of course, means minus, which means less. Energy subtracted. Less energy, now, in the black hole. 'Hawking radiation' is the radiation (the positive-energy particles that escape to a distance) emitted by black holes in this process.

Another more mind-boggling way of looking at this process is to regard the particle that falls into the black hole as a particle that is really travelling backward in time. One characteristic of the underlying physics of the universe is that with very few exceptions physical reactions work equally well backwards and forwards in time. There is no preferred direction. For most of us, accustomed as we are to living with an arrow of time that always points from the past to the future, this is not easy to accept, but it is something physicists have got used to. So it isn't completely off-the-wall to think of the particle falling into the black hole as a particle coming out of the black hole, but travelling backward in time. When it arrives at the point in spacetime at which it first materialized, the gravitational field scatters it so that it travels forward in time.

Yet another way of viewing all this – which would seem to bend the laws of physics even more than having something come out of a black hole – is to note that the uncertainty principle makes it possible for photons to travel at slightly more than the speed of light over extremely short distances. How can this be? An 'extremely short distance' implies that we are able to measure the position of the particle with great accuracy. We know precisely where it is. Recall the Heisenberg uncertainty principle and the see-saw. If we know exactly where a particle is, we can't simultaneously know exactly how it is moving. So the speed at which this particle is travelling becomes something we cannot state precisely. Saying that it travels at the speed of light *would* be a very precise statement. Instead, we must allow a little leeway and say that over extremely short distances, it is possible for a photon (which normally travels at the speed of light) to travel at slightly more, or less, than the speed of light. That means it's possible for a photon to escape from a black hole. This way of looking at Hawking radiation makes it easier for us to understand why more particles escape from smaller black holes than from larger ones. The thickness of the barrier around a black hole is proportional to the size of the hole, so a particle would have to travel a greater distance at greater-than-light speed to get through the thicker barrier of a large black hole.

Whichever of these explanations is most meaningful to you, and I hope one of them is, it is clear that with the discovery of this radiation (Hawking's second famous discovery about black holes), he showed that his first famous discovery, the Second Law of Black Hole Dynamics (that the area of the event horizon can never decrease) does not always hold. In the Hawking Process, a black hole may get smaller and eventually evaporate entirely.

How does Hawking radiation result in the event horizon decreasing and the black hole getting smaller?

You learned very early in this book that any change in the mass of a body changes the amount of gravitational pull it exerts on another body. If the earth were to become less massive (not just smaller, as in the shrinking earth story, but less massive), its gravitational pull would be less strong out at the radius where the moon orbits. If a black hole becomes less massive, its gravitation pull will be less strong where the event horizon has been. The escape velocity there becomes less than the speed of light. The curvature of spacetime there is no longer great enough to cause photons to hover there and not allow them to escape. That ceases to be the event horizon. There is a new event horizon at a smaller circumference. This is the only process we known in which a black hole can get smaller.

Getting back, then, to what we can measure about the particular black hole we've been investigating: why not measure the Hawking radiation coming out? We'll be disappointed. From a black hole this size, very few particles can escape. It will have a surface temperature of less than a millionth of a degree above absolute zero. Hawking tells us, 'A ten-solar-mass black hole might emit a few thousand photons a second, but they would have a wavelength the size of the black hole and so little energy we would not be able to detect them.' Our 100-solar-mass black hole is larger than that, so it will emit even less. Here's the way it works: the greater the mass, the greater the area of the event horizon. The greater the area of the event horizon, the greater the entropy. The greater the entropy, the *lower* the surface temperature and the rate of emission.

With such a fizzle as that, why all this fuss about Hawking radiation? Because in some instances it isn't a fizzle. Hawking suggested in 1971 (before he discovered Hawking radiation) that there are black holes which did not form in the way we've described in this book. These were never stars. He calls them primordial black holes because, if they exist, they are relics of the early universe when there were pressures that could compress even a small amount of matter – far less than the Chandrasekhar limit – sufficiently to form a black hole. The most interesting ones, according to Hawking, are those about the size of the nucleus of an atom or smaller. Whatever size it started out, a primordial hole has been losing mass and getting smaller for a very long time. The smaller a black hole is, the easier it becomes for particles to escape and the hotter its surface temperature. A primordial black hole would crackle with radiation. 'Such holes', says Hawking, 'hardly deserve [to be called] *black*: they really are *white hot*.'

The consequences are drastic for the hole. As its mass and size decrease, its surface temperature and rate of emission of particles increase. The white-hot black hole loses mass more and more rapidly, the lower the mass, the higher the temperature – a vicious circle. How the story ends is still a matter of speculation, but Hawking's guess is that the hole will disappear in an enormous final puff of particle emission, like millions of hydrogen bombs exploding.

Our 100-solar-mass black hole is not going to do that any time soon. At the rate this hole emits particles, according to Hawking there is a chance the universe will end long before it reaches that stage. In fact, there's also a good chance enough matter will continue to fall into the hole to more than make up for the mass it is losing through Hawking radiation, and it won't grow smaller at all.

What can we learn?

There are other experiments to run. We'll want to continue throwing a variety of items into the hole – test objects of different sizes, masses and electrical charges. Some of these should be microscopic test particles. We'll send them down at many different angles. As we watch these items approach the hole and disappear, we can study gravity waves and other radiation that result. Similar experiments can teach us more about the rotation of the hole, how that rotation affects space nearby, the ergosphere, tidal effects, time dilation, and redshift.

For example, watching a small object fall toward the black hole, we know that, according to theory, somewhere still outside the event horizon the object should begin to emit gravity waves more and more strongly (Figure 6.8). As the test object approaches the event horizon, we notice the frequency and energy of the gravity waves greatly reduced by redshift. We won't observe any emission from the object after it reaches the event horizon.

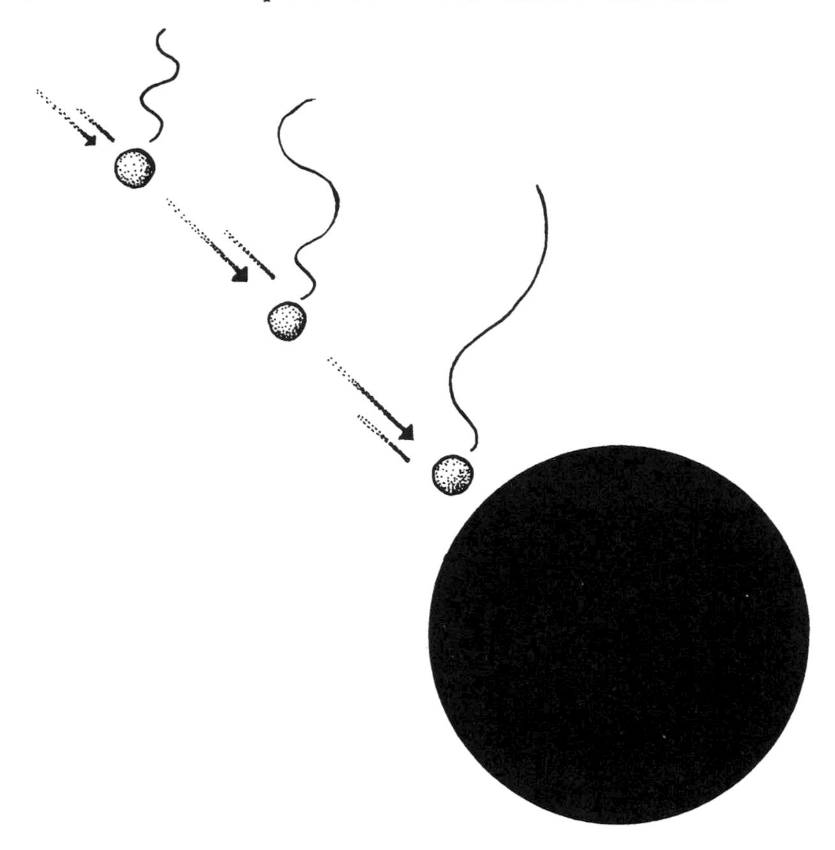

Figure 6.8.

If the hole is rotating, even test items sent straight down from the ship may end up circling the hole many times before falling in, and we shall begin to understand how 'accretion disks' might form around black holes. We'll get to those in more detail in the next chapter. By way of introduction: when tidal effects stretch and tear apart matter approaching the event horizon of a black hole, they eventually reduce it to a form of gas. If there is a great deal of gas, there are violent collisions and a great deal of friction among the molecules in the gas as they move nearer and nearer to the hole. Imagine a great volume of traffic on motorways and smaller roads converging from all directions on one destination. Imagine further that even after there have been enough collisions, fender-benders, and slow crunches to crush the vehicles to powder, whatever is left of the vehicles is somehow still able to continue moving toward that central destination. This analogy gives us some idea of the situation as matter is drawn into a black hole.

Around the equator of a rotating black hole, the spiralling gas forms a disk – an 'accretion disk' – that moves around the hole like a bloated compact disk in a CD player. The friction between the molecules in the accretion disk causes everything to heat up. The disk emits a great deal of radiation. The closer to its centre, the higher the energy of that radiation. We're going to see in Chapter 7 that one of the ways we have of identifying black hole candidates in space is to look for evidence of such radiation.

While the black hole affects things in the area of space around it, it in turn is affected by nearby objects. The area of the event horizon and the size of the black hole will increase as things come too close and are drawn in. We may discover that the gravitational pull of other massive bodies raises tides on the event horizon, and we may find that this slows the hole's rotation. Matter falling in may also change the speed of rotation.

Penrose Potshots

Riding the train into London on an afternoon in 1969, on his way to lecture at the University of London, Roger Penrose came up with an idea that soon became known as the Penrose Process. In the Penrose Process two particles perform an exchange of energy within the ergosphere of a rotating black hole. As with Hawking radiation (which happens at the event horizon, not in the ergosphere, and also occurs when the hole is not rotating), one of the particles acquires a debt of energy to the other and then makes an escape to a distance, never to repay the debt, while the other particle falls into the black hole, in effect transferring its loss to the black hole. The hole ends up with less energy of rotation and less speed of spin.

Some of the sporting members among our passengers and crew may want to try their hands in a test of cosmic marksmanship. The first object of the game we shall call Penrose Potshots is to slow down the hole's rotation by placing particles with precisely the right speed and direction in the ergosphere. As an escaping particle carries away some of the angular momentum and mass/energy of the black hole (as negative energy goes into the hole with the cheated particle), the rotation slows down. Eventually the hole might have no angular momentum left at all and stop spinning. However, as we mentioned earlier, a black hole has a certain irreducible mass below which it cannot go in this process. Think of it this way: a ball that is spinning has greater energy (hence, also greater mass – remember Einstein's equation) than an identical ball at rest. Stopping its spin entirely reduces its mass/energy, but it cannot reduce it to less than the ball at rest. Likewise, the Penrose Process cannot reduce the mass/energy of a black hole below the level it would be if the hole were not spinning at all, while the Hawking Process can.

The second object of Penrose Potshots is to spin up the hole again by shooting particles into it off centre, giving back to it the angular momentum that was carried away. The restored spin and some particles that fall in will also of course restore mass/energy to the black hole.

Extracting mass/energy and angular momentum from a black hole and then getting the hole back exactly to its original mass/energy and angular momentum is by no means child's play. This will be a true test of skill in calculation and marksmanship. We'll discover that if we restore all the angular momentum we may find ourselves with more than thc original mass/energy. But with precisely correct choices of particle masses, direction of motion and impact points we may manage to reverse exactly the changes we have made.

A test of time

A last experiment. We'll position extremely accurate clocks about at various distances from the hole, compare them, and test theories about time dilation. We'll find that time comes to a stop (from our ship-at-a-distance's point of view) at the event horizon. How tempting to build a platform just outside this hole's event horizon and live there, prolonging our lives, maybe becoming immortal!

From one way of looking at it – yes, we could do that. Not too comfortably, certainly, with the tidal effects we would experience there. But surely you understand by now that our lives on that platform would not seem any longer. To us, seconds, minutes, hours, days, years, decades would pass as they do now. We couldn't know how slow our time was moving or appreciate our immortality unless we could see ourselves from somewhere else in the universe. Of course, if we were to return to earth just a year later by our time, we would find that billions of years – perhaps almost the entire future history of the universe – had already flown by there.

7

Evidence in the case

Artist's conception of Cygnus X-1 (left) and its optical companion star (right). Matter pulled off the optical star forms an accretion disk as it flows into the (suspected) black hole. (Air-brush drawing by Lola Chaisson.)

If it isn't a black hole, it really *has to be something exotic!*
Stephen Hawking

When Wheeler coined the name 'black hole' in 1967, there was no evidence from observational astronomy pushing anyone to believe such phenomena really exist. In fact, before 1964 there hadn't even been a serious suggestion about what evidence to look for.

Occasionally in science, albeit rather rarely, a theory emerges that compels us to believe it is correct simply because it makes such elegant mathematical sense, even though there is no experimental or observational evidence to back it up. In the case of black holes, the more physicists and mathematicians toyed with the idea, the more beautifully logical it seemed. The more they fought it, the more it became unavoidable. Certainly no one could prove it *wasn't* true – but it was still 'only a theory'. Hawking comments, 'I do not know any other example in science where such a great extrapolation was successfully made solely on the basis of thought. It shows the remarkable power and depth of Einstein's theory.'

By the mid-1960s, the solutions physicists had discovered to Einstein's equations made it difficult not to conclude that black holes must exist, but in the absence of observational evidence many scientists and mathematicians were slow to capitulate. It was intriguing to speculate about what would happen to a star too massive to become a white dwarf or neutron star, and it made for fiendish mathematical exercises to assign to graduate students, but taking black holes seriously as something that might really be out there – that was another matter.

Nineteen sixty-seven was a watershed year when it came to attitudes about black holes. The incident that triggered the shift was the sort of stuff science fiction fantasies are made of. Jocelyn Bell, a graduate research student of Antony Hewish at the University of Cambridge, using receivers set up like rows of beanpoles in a field near Cambridge, detected mysterious, regular pulses of radio waves coming from space. The discovery was chilling, because it seemed this might be our first contact with an alien civilization. It was not, though Bell and Hewish only half in jest dubbed the radio sources LGM for Little Green Men. What decided the case against the aliens was that the sources of the pulses were not moving in patterns like orbits. We would expect a civilization to live on a planet orbiting a star, not on a star. A more promising explanation was that the pulses were coming from a star rotating on its axis at a furious pace many times a second and sending out a narrow beam of radiation across space. Like the beacon of a hyperactive lighthouse, the beam would sweep the earth with each rotation of the star. Pulsars, as these stars were dubbed after it was clear they were not Little Green Men, turned out to be tiny, extremely dense stars only about 30 kilometres (20 miles) across. Having read previous chapters of this book, you will perhaps find that description familiar. Bell had discovered a neutron star.

If this was a let-down for those hoping to discover extraterrestrial life, it was tremendous encouragement for those hoping to find black holes. Many astronomers had been clinging to the notion that all massive stars in their dotage and death throes lose enough matter to bring their masses safely below the Chandrasekhar limit – all ending up as white dwarfs. But if some stars retain enough mass to collapse to the size and density of neutron stars, was it so absurd to think that others would end up more massive yet and form black holes?

Since 1967, astronomers have discovered many neutron stars. It took longer to find convincing evidence of a black hole, and meanwhile some scepticism lingered. Can nothing halt the collapse of a star after gravity overcomes the exclusion principle? Perhaps there is another opponent. Who can say what happens within the event horizon? Can we so completely trust these solutions to Einstein's equations? General relativity tells us that mass curves spacetime and the larger and more concentrated the mass, the greater the curving. We know this has been tested in less drastic situations but we also know that in the ultimate drastic situation, where spacetime curvature approaches infinite, the theory itself breaks down. Does the collapse somehow come to a halt before the star reaches a point of infinite density? Penrose had said no, but the Heisenberg uncertainty principle probably disallows a point of infinite density – an extremely precise measurement of position and momentum! Taking that into account, we refined 'infinite' to read 'near infinite'. Hedging our bets? On the other hand, maybe gravity self-destructs. If it swallows light, does it also swallow itself? No one was claiming to have figured out quantum gravity. Would something in those yet-to-be-discovered laws halt the collapse? (Keep in mind, however, that not having a singularity of infinite density, or even near infinite density, doesn't rule out having a black hole. Once the event horizon forms, even if the collapse were to stop just beyond it, we have a black hole.)

All things considered, surely to the uninitiated it seems nonsensical that something many times more massive than our sun would collapse to an infinitely small point. We cannot visualize this happening even in our wildest imaginations. Both the largeness and the smallness of it defeat us. Nevertheless, in the late 1960s, the 1970s and the 1980s it became increasingly the case that if someone were to have proved that black holes don't exist, it would have caused an upheaval of awesome proportions in the world of physics and sent some of our finest mathematicians and physicists back to square one! This, in spite of the fact that observational evidence didn't become overwhelmingly convincing until the 1990s.

When it comes to discovering real black holes out in space, we have come a long way since the late 1960s. At first we observed things going on out there that led us to speculate, 'There might be a black hole there causing that.' As telescopes improved, Hawking felt justified in commenting about some observational evidence, 'If it isn't a black hole, it *really* has to be something exotic.' In other words, of all possible explanations for this, a black hole is actually the least difficult to believe. In the late 1980s and the 1990s, researchers have at last detected evidence that causes us to say 'There seems *no* other sensible way to explain this than by concluding there is a black hole there.' But the culprit itself always remains out of sight. So, of course, does the wind, though we know it's there. As we have said before, evidence of black holes always comes in the form of circumstantial evidence, because no telescope can show us a picture of a black hole in person.

It's time to look at this circumstantial evidence. In this chapter and the next we will focus on the discoveries of black holes circling in binary star systems in our galaxy and of much more massive and powerful black holes in the cores of other galaxies and quasars. In Chapter 9 we'll explore the possibility of a black hole lurking in the heart of our galaxy and follow other highly promising leads. Since black hole evidence comes in the form of radiation in every part of the electromagnetic spectrum, it would be well to keep Figure 3.1 close at hand.

With earlier chapters in mind, you may wish to second-guess the experts and make your own list of what to look for in this quest. Your list will surely include such things as lensing effects, accretion disks, gravity waves, certain patterns of radiation, unusual movement of objects in space. It was with the last item on that list that the search for real black holes in space became a serious though not widespread endeavour even before Jocelyn Bell's discovery of a neutron star.

Wheeler's 'Great Waltz'

Imagine a dimly lit ballroom with the women all dressed in white gowns and the men in black formal wear. In the semi-darkness, only the women are visible as they whirl around in the arms of their companions. This picture is John Wheeler's way of introducing the possibility of a black hole in a binary star system.

A binary system consists of two stars linked by each other's gravity, moving around one another like the men and women in Wheeler's ballroom. It isn't really correct to think of them orbiting around one another. Instead, they orbit around their common centre of mass. The centre of mass is an imaginary point lying between the two stars – 'imaginary' in the same sense that the equator is an imaginary line or an event horizon is an imaginary surface. There is nothing really there to be seen. The centre of mass is always closer to the more massive star. Think of two children, one much heavier than the other, riding an adjustable see-saw. If we want them to balance, we adjust the fulcrum of the see-saw so that it is much nearer the heavier child. When the see-saw is moving, during the ride, the heavier child won't move as much as the lighter child. By the same principle, in Wheeler's ballroom, the heavier man will not move as much as the lighter woman. In a binary system the more massive star is nearer to the centre of mass and will not move as much as the less massive star. If the discrepancy in mass is very large, the difference between their movement will also be very large.

Our sun has no partner in Wheeler's ballroom, but not all stars are wallflowers. There are many binary systems in our galaxy in which we are able to see both partners – both 'dressed in white'. Of more interest in the context of this chapter: what if we observe, as Wheeler's analogy suggests, instead of two stars in what is obviously a binary system, only one, moving as though it were in orbit with a partner but with no partner visible? We would rightly suspect that a star behaving like that isn't alone out there. This woman in white does have a partner. A black hole? Perhaps, but not necessarily. The companion might also be a small, dim, low-temperature star, a white dwarf, or a neutron star.

From an observational point of view, the most significant clue in the search for these 'invisible' companions has turned out not to be simply finding a star waltzing around as though it were only pretending to have a partner, but the detection of very luminous, rapidly flickering X-ray sources. It doesn't quite make sense to call these companions 'invisible'. Perhaps Wheeler would allow us to say that, viewing the ballroom through an X-ray telescope rather than an optical telescope, we discover the dark-suited men are carrying sparklers, while the women hardly show up at all! As we proceed we shall more properly refer to the 'invisible' partner in a binary system as the 'compact star' or the 'compact companion star'.

X-rays are considered a hallmark of accretion of matter onto a compact star. Black holes and neutron stars – both of which are 'compact stars' – have what physicists call deep 'gravitational wells'. Material falling onto such a star liberates a lot of gravitational potential energy. Carole Haswell, an astronomer at Columbia University whose work we shall discuss later, asks us to think of it like this: A brick dropped from the roof of a skyscraper gains more energy (gravitational potential energy) during its fall than would the same brick dropped from the roof of a single storey building. The more compact the star the higher the skyscraper we must imagine in this analogy . . . and (in the jargon) the deeper the gravitational well. When we see X-rays, which you know from the diagram of the electromagnetic spectrum (Figure 3.1) are high-energy photons, we conclude that there must be a source of considerable energy powering their emission, and dropping material toward a black hole or neutron star is one of the best ways we know to liberate energy. In fact, more energy can be liberated for a given amount of mass than would be liberated by a hydrogen bomb of the same mass.

However, while a great outpouring of X-rays, flickering rapidly, makes a good case for the presence of a compact star, it doesn't necessarily spell out 'black hole'. In order to say that the compact star is a black hole, it is necessary to establish that its mass is too great for it to be a neutron star. We saw in Chapter 1 that if it is more than about three solar masses it is unlikely to be a neutron star – it must be a black hole. How do we establish its mass?

In Wheeler's ballroom, we can make a good guess as to the mass of one of the invisible men by watching the speed of the woman and noting how large she is and the size of the circle in which she swirls. By similar reasoning, we are able to get a fairly good idea of the mass of compact companion stars (the 'invisible' partners) in binary systems. Here, roughly, is the rationale involved. The more massive the compact star is, the stronger its gravitational pull is on its visible partner. The greater that gravitational pull, the more centrifugal force the visible partner needs to keep from being pulled toward the compact star. The way to acquire sufficient centrifugal force to resist the pull of a very massive compact companion star is to move rapidly in orbit. The more massive the compact star, the more rapid the movement of its visible partner will have to be. If we want to find out whether the compact star is a black hole, it stands to reason that we might find the orbital velocity and mass of its visible partner, and from that infer the compact star's mass. If the compact star is sufficiently massive, more than about three solar masses, then it is probably a black hole, not a neutron star – all of which sounds nicely cut and dried. Unfortunately, we are going to find out that it isn't all quite so neat.

Figure 7.1. Yakov Borisovich Zel'dovich, the great Soviet theoretical physicist and astrophysicist, mentor for generations of Soviet astrophysicists, at the chalkboard in his apartment in Moscow. (Photo taken by Kip Thorne.)

Theorist Yakov Borisovich Zel'dovich (Figure 7.1) and his astronomy graduate student Oktay Guseinov at the Institute of Applied Mathematics in Moscow were thinking along these lines back in 1964 when they pioneered the search for real black holes by combing the many hundreds of listings of binary systems that astronomers had already observed and catalogued. This was the same Zel'dovich who was later to trigger Hawking's discovery of Hawking radiation – see Chapter 6. Zel'dovich and Guseinov came up with five candidates they thought were too massive and compact to be neutron stars.

We have said that one piece of information that can help us

calculate the mass of a compact star, as Zel'dovich and Guseinov were trying to do, is the 'orbital velocity' of its visible partner – that is, how rapidly the visible partner is travelling as it orbits. How can we find out a star's orbital velocity? On first thought, the most obvious method would be to measure it the way we normally measure speed, from the distance travelled (the size of the orbit) and the length of time it takes to go that distance (the 'orbital period'). Unfortunately, it isn't that simple. We are forgetting that astronomers are not watching a little shiny ball circling in an orbit. They are looking at a pinpoint of light. However, there is an indirect way of getting at the orbital velocity.

Recall our discussion of the Doppler effect and red and blue shift; then imagine two stars circling one another – one we can see, one we can't. We will ignore for the moment the one we can't see. As the visible star travels toward us in its orbit, the Doppler effect shifts that star's light toward the blue end of the spectrum. As it travels away from us, the shift is toward the red end. By measuring the shifts in wavelength, it is possible to figure out the speed at which the star is moving toward us or away from us and how this velocity changes over time. So far, so good. However, there is a hitch. What this method does not allow us to measure is the speed the star is meanwhile travelling *perpendicular* to us. For that reason, measuring the speed it travels toward or away from us can't be counted on to give us the actual orbital velocity of the star. Instead, it gives us the so-called 'line-of-sight' velocity (or 'radial' velocity).

Astronomers know that the line-of-sight velocity is the *minimum* orbital velocity for this star, the *lower limit* to what it could possibly be. They are able to say that the star's actual orbital velocity is surely this great or greater, it cannot be less. Saying this in the case of black hole candidates is sometimes saying a lot, as we shall see in a moment, but it would be very nice to be able to say more! However, the line-of-sight velocity gives us the actual orbital velocity (they are one and the same) only if we happen to be viewing the binary system 'edge-on' (as we view a record turning on a phonograph turntable if we sit with our eye level with the plane of the record). Unfortunately, from earth we view very few binary systems edge-on – only one that we know of, in fact, that promises to contain a black hole. We are aware that we're viewing a binary system edge-on when the stars take turns eclipsing one another. If they are not doing that, then we're not viewing the system edge-on and the actual orbital velocity of the star is indeed greater than the line-of-sight velocity (the lower limit) we've measured. How much greater? If we knew how far from edge-on we are viewing the system, we would know that answer. But there's the rub. It is not at all easy to determine how much a binary system is tipped. In fact, though there are some ideas on offer, we don't yet know how to find out.

Dead end? Not quite, but this does leave us a little adrift if we are trying to figure out how massive the compact companion star is by measuring the speed at which its visible partner orbits! All we can know with certainty is the minimum that speed can be, so all we can know with certainty about the compact companion's mass is the minimum that mass can be. If that minimum is clearly above 3 solar masses, well and good. We need say no more to make our case for a black hole. Of course the actual orbital velocity of the visible partner might be much larger than the line-of-sight velocity, and so might the mass of its compact companion. Thus, even if the minimum mass is too small for a black hole, that doesn't rule out the actual mass being large enough. It only rules out our knowing for sure whether it is.

As you continue to read this chapter, you may wonder why we see such a wide range of masses given for some compact stars (in the case of Cygnus X-1, for example: 'A minimum of 3 solar masses, probably greater than 7 solar masses, and most likely about 16 solar masses'). Why can't these astronomers get their act together? We have just seen part of the explanation. Other uncertainty comes from the difficulty of discerning the mass of the visible star.

To sum up the predicament astronomers find themselves in when they try to discover whether a compact star is a black hole: four key measurements are involved in calculating the mass of the compact companion star in a binary system.

1. The orbital period (how long the two stars take to complete

one orbit). Measuring the red and blue shifts of the visible star allows us to calculate this without difficulty.

2. The line-of-sight velocity (also known as radial velocity) of the visible star. We've seen that this also can be calculated from the red and blue shifts and gives us the minimum orbital velocity for the star – or the actual orbital velocity *if* we are viewing the system edge-on.
3. The mass of the visible star.
4. How close to edge-on we are viewing the system.

The first two measurements – orbital period for the system and line-of-sight velocity for the visible star – *can* be made precisely. The last two, in most cases, so far, cannot.

In 1966, two years after he had begun searching for massive compact companion stars in binary systems, Zel'dovich and a colleague Igor Novikov hit upon the notion of combining the binary system idea with a search for X-ray emission. As they saw it, identifying strong black hole candidates would require using both optical telescopes *and* X-ray detectors. A promising candidate would be in a binary system in which one partner shows up brightly in the visible part of the spectrum but is dark in the X-ray part of the spectrum; while the other partner is dark in the visible part of the spectrum but bright in the X-ray part. Zel'dovich and Novikov led the way in thinking about how such X-rays would be produced as gas streams from the visible star toward the neutron star or black hole. Then in 1969 Donald Lynden-Bell of Cambridge improved on previous understanding of how the gas would move and coined the term 'accretion disk'.

To elaborate upon what we have said earlier about accretion disks: gas flowing off the visible star's surface is drawn toward the black hole or neutron star. As it falls it converts its gravitational potential energy into kinetic energy (energy having to do with its motion). Instead of falling directly into the black hole or onto the surface of the neutron star, the gas forms a disk like a bloated CD around the compact star. In the disk, collisions and friction (recall our analogy of converging traffic) between the atoms of gas convert the energy of its orbital motion into heat energy. The gas loses altitude with respect to the black hole or neutron star, spiralling lower and lower, moving toward the centre of the disk. As the gas approaches the inner edge of the disk, the edge near the black hole or neutron star, the temperature of the gas rises prodigiously. It is heat energy from the disk that we see radiated as X-rays.

In 1966, when Zel'dovich and Novikov first suggested that X-ray emission would be a clue in the search for black holes, X-ray astronomy was not highly enough developed to make a search for X-ray emission from binary systems practical. To put things in historical perspective: the first 'X-ray star' had been discovered in 1962 with an X-ray detector flown on a rocket above the earth's atmosphere in a project to measure X-rays coming from the moon. (X-rays don't penetrate the earth's atmosphere.) Astronomers then lacked theories such as Zel'dovich and Novikov's to convince them that the strong X-ray emitter they had found might be a neutron star or black hole. The first positive evidence of the existence of neutron stars – the pulses detected by Bell and Hewish in 1967 – came not from an X-ray detector but from a radio telescope. By the early 1970s there was substantial improvement in X-ray detectors. The first X-ray satellite, Uhuru, was launched in 1970. Uhuru discovered and catalogued 339 X-ray stars within two years. The Einstein telescope, launched in 1978, was the first real X-ray telescope; that is, the first telescope capable of focusing X-rays to make images of the X-ray sky, as an optical telescope gives us images of the 'visible' sky.

Cygnus X-1

One of the strongest black hole candidates these instruments have discovered and studied with increasing precision is in the area of the sky known since ancient times as Cygnus, the Swan. It is Cygnus X-1, or Cyg X-1 for short. 'Cygnus' designates the constellation; 'X' indicates that it is an X-ray source; '1' tells us that it is the brightest in that constellation.

Cygnus X-1 was the subject of a now famous bet between Stephen Hawking and Kip Thorne (Figure 7.2). In 1974, ten years after Cygnus X-1 was first observed by a detector in a rocket flight and three years after it was identified as a strong black hole candidate, the evidence still left experts only about 80% certain that it was a black hole. That December, Hawking agreed that if Cyg X-1 turned out to be a black hole he would give Thorne a one-year subscription to the magazine *Penthouse*, while Thorne agreed that if Cyg X-1 turned out not to be a black hole, he would give Hawking a four-year subscription to *Private Eye*. Hawking justified his puzzling bet against the black hole explanation by calling it 'insurance'. As he put it, 'I have done a lot of work on black holes, and it would all be wasted if it turned out that black holes do not exist. But in that case, I would have the consolation of winning my bet'.

Before finding out who won, let us take a look at the evidence for ourselves and perhaps place a few side wagers.

Cyg X-1 is the sort of black hole candidate Zel'dovich had been looking for. Here is a binary system in which an optically bright but X-ray dark star orbits with an optically dark but X-ray bright compact star (Figures 7.3 and 7.4). This binary system is in our galaxy, about 6000 light years from earth. With an optical telescope we see a star that seems to be a blue giant star. It's too dim to be detected with the naked eye but bright in comparison with most stars seen with a large optical telescope. Studies of the Doppler shift in its light show us that this visible star must have a companion star, though that star can't be seen with optical telescopes, and that the two stars complete one orbit in 5.6 days.

Cygnus X-1, the blue giant's partner, though dark in the part of the spectrum we can view with an optical telescope, is one of the brightest objects in the X-ray sky. Studies of the X-ray radiation tell us that gas heats to perhaps a hundred million degrees Celsius in the accretion disk. (See the Frontispiece of this chapter.) The X-ray emission is not only strong but fluctuates violently and chaotically, as we expect from the accretion of matter onto a black hole or neutron star. Cygnus X-1's mass is at a minimum 3 solar masses, probably greater than 7 solar masses, and most likely about 16 solar masses.

In the years since 1974, 80% certainty that Cygnus X-1 is a black hole has increased to about 95% certainty. Why not 100%?

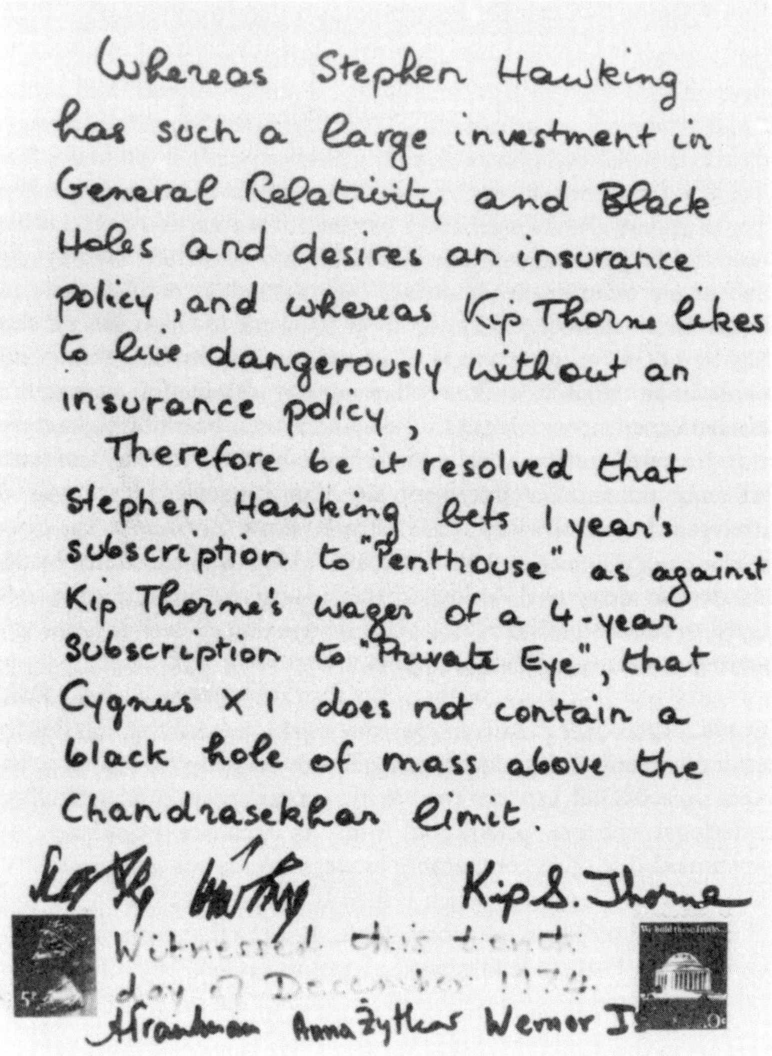

Whereas Stephen Hawking has such a large investment in General Relativity and Black Holes and desires an insurance policy, and whereas Kip Thorne likes to live dangerously without an insurance policy,

Therefore be it resolved that Stephen Hawking bets 1 year's subscription to "Penthouse" as against Kip Thorne's wager of a 4-year subscription to "Private Eye", that Cygnus X 1 does not contain a black hole of mass above the Chandrasekhar limit.

[illegible] Kip S. Thorne

Witnessed this tenth day of December 1974.

[illegible] Anna Żytkow Werner I[illegible]

Figure 7.2. The certificate documenting Stephen Hawking and Kip Thorne's bet about whether Cygnus X-1 is a black hole. (Courtesy of Stephen Hawking.)

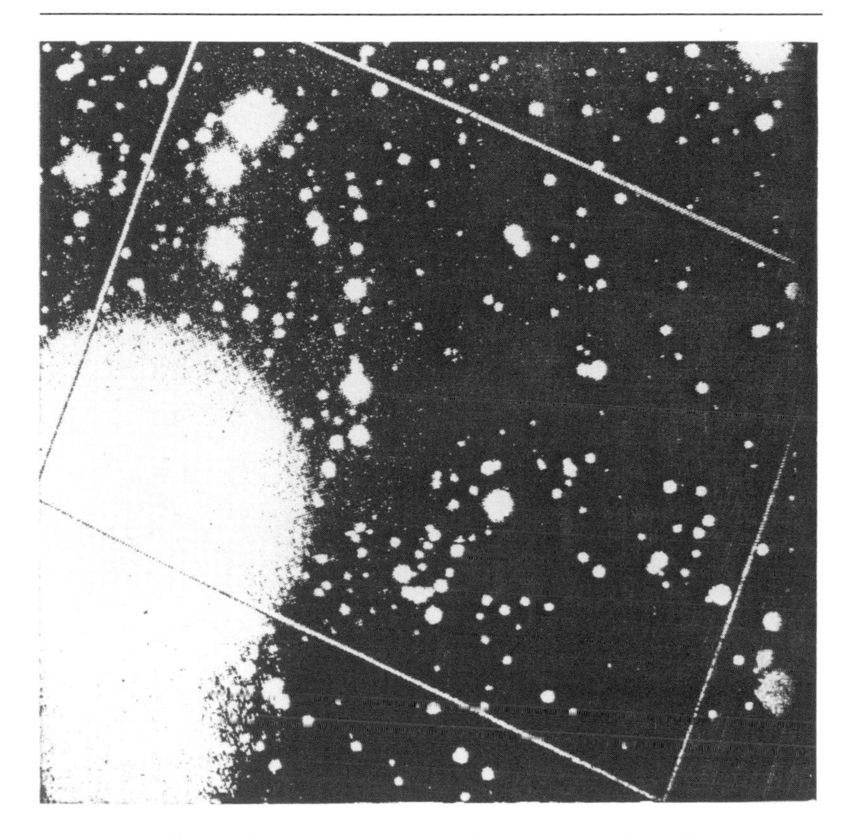

Figure 7.3. A blue giant star, the visible companion of Cygnus X-1. The rectangle frames the region where the black hole candidate Cygnus X-1 is known to be lurking. (Harvard-Smithsonian Center for Astrophysics.)

For one thing, Cygnus X-1 has a minimum mass close to the 3-solar-mass dividing line between neutron stars and black holes. This dividing line is not clear-cut and is still somewhat controversial. No one is yet certain, for instance, how much or how little centrifugal force may affect the collapse of a borderline star. Of course, we don't really know that Cygnus X-1 *is* a borderline star, only that the lower limit of its mass is at the border. But without a precise knowledge of how far the system is tipped we can't say unequivocally that its mass is actually well above 3 solar masses. On another front, there are those who insist that the mass of the

Figure 7.4. X-rays coming from the mysterious Cygnus X-1 source. What you see in this photograph is not an object but X-ray radiation coming from the area within the rectangle in Figure 7.3. The image was obtained by the Einstein Observatory, a spacecraft outside the earth's atmosphere. (Fred Seward, Smithsonian Astrophysical Observatory.)

visible star, the blue giant, has been grossly overestimated, an error which could throw off considerably the calculations of Cygnus X-1's mass.

What have Hawking and Thorne made of all this, and what has happened to their bet? Hawking has decided that the evidence gives us better than 95% certainty that Cygnus X-1 is a black hole. In June 1990, when he was lecturing at the University of Southern California and Thorne was away in Moscow, Hawking, aided and abetted by friends, broke into Thorne's office, removed the bet document from its frame, and wrote a note on it conceding the bet. Hawking's thumbprint serves as his signature.

Even if Cygnus X-1 should turn out not to be a black hole, that needn't mean, as Hawking seemed to fear when he took out his 'insurance policy', that there are no such things as black holes.

Among the compact stars in binary systems in our galaxy there are stronger candidates than Cygnus X-1.

A0620-00 Monoceros

A0620-00 is in the constellation Monoceros near Orion, about three thousand light years from earth. Light from its accretion disk is visible to optical telescopes, though not light from the star itself.

A0620-00 first called attention to itself in an intense X-ray outburst detected by the British satellite Ariel V in 1975 (Figure 7.5). Searching the archives, astronomers found that A0620-00 had erupted similarly early in the century, in November 1917, but no one had given it much notice back then. We now know that with both neutron stars and black hole candidates, X-ray emissions and sometimes emissions in the optical part of the spectrum as well flare up and fade randomly over periods of seconds or even years. (In the case of black holes, these emissions must of course come from the accretion disk, not the black hole itself.) The Ariel V seems to have caught A0620-00 in one of its relatively rare 'switched on' phases.

In some of the strongest black hole candidates, A0620-00 among them, there is a distinctive pattern behind the random fluctuation, which sets these candidates apart from many other X-ray emitters in the sky: a large flux of X-rays in the low-energy end of the X-ray spectrum (review Figure 3.1) that trails off into a tail of X-rays from the high-energy end, even over the border into gamma-ray range. Could this pattern be the give-away black hole fingerprint that astronomers would so like to find? '[Why would] the absence of a surface change the spectra in such a way that it gives you soft X-rays with a hard X-ray tail?' asks Charles Bailyn of Yale University, who has chosen to study another candidate showing the same pattern, Nova Muscae. Why indeed couldn't the same pattern emanate from a neutron star? So far there are unfortunately no answers to those questions, no theoretical model to boost confidence that this pattern really is a black hole finger-

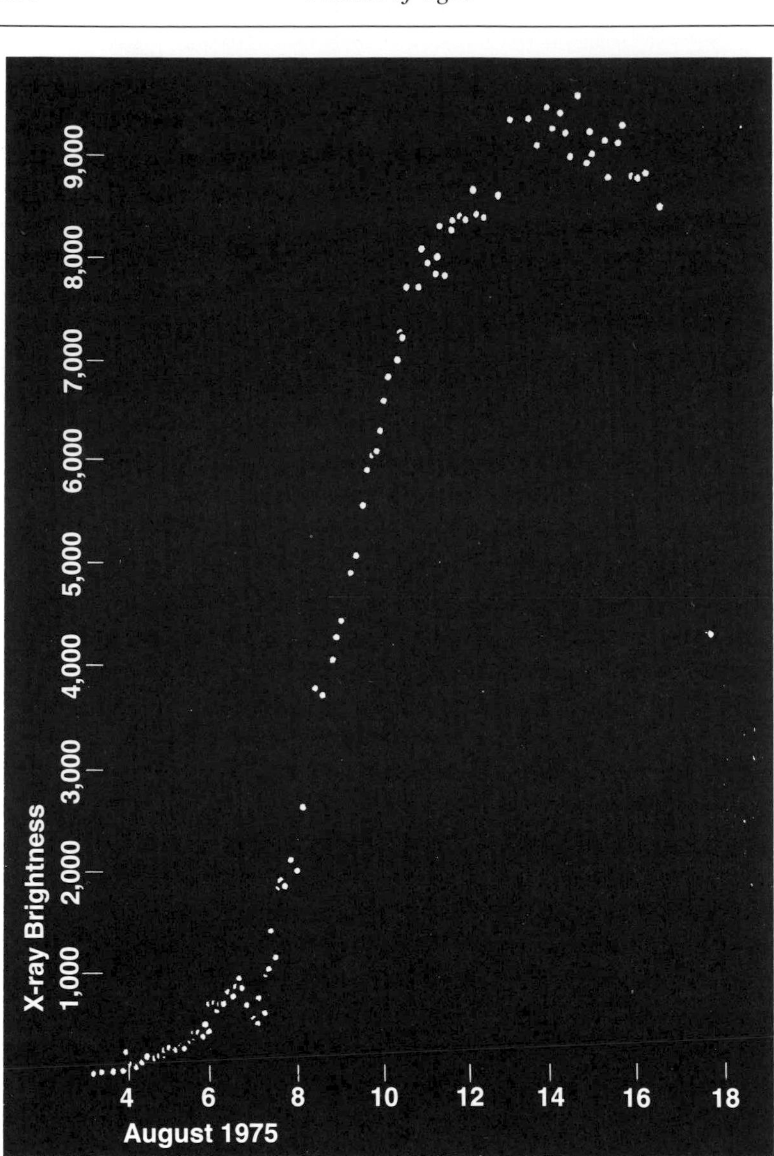

Figure 7.5. A0620-00 called attention to itself in 1975 with this X-ray outburst, noted by astronomers using the Ariel V X-ray satellite. The numbers across the bottom of the picture indicate dates in August 1975. An optical outburst followed the peak intensity of the X-rays. (From data by University of Leicester X-ray Astronomy Group, courtesy of Kenneth Pounds.)

print. However, there are other good reasons to conclude that A0620-00 is a black hole.

Jeffrey McClintock of the Harvard-Smithsonian Centre for Astrophysics and Ronald Remillard of MIT were the first to discover how strong a candidate A0620-00 is. In 1986, they measured the shifts in the spectrum of its visible partner, an orange dwarf, and found that the orange dwarf's orbital period is 7.75 hours and that its line-of-sight velocity is 457 km per second. (For comparison, the earth orbits the sun at only 30 km per second.) McClintock and Remillard deduced that A0620-00 – the dwarf's compact companion – must have a mass of at least 3.2 solar masses. Of course, the mass is much greater than this if we are not viewing the system edge on, and we are very probably *not* viewing this system edge on. When it comes to the other problematic measurement – the mass of the visible star – A0620-00 has an advantage. The orange dwarf is simply too small to make a difference. That mass is largely irrelevant.

There have been some interesting glitches in the study of A0620-00 that point up the difficulty of studying a black hole candidate. In 1990, Carole Haswell, then of the University of Texas, and Allen Shafter of San Diego State University studied the shifts in the spectrum of light coming from A0620-00's visible accretion disk. They found that A0620-00 completes a revolution around the centre of mass every 7.75 hours (which, as expected, accorded with the time clocked by its orange dwarf companion as noted by McClintock and Remillard), attaining a line-of-sight velocity of about 43 km per second. From those numbers Haswell and Shafter calculated the circumference of the orbit and the radius of the orbit – which is the distance from the black hole to the system's centre of mass – and estimated the mass of A0620-00 at 3.8 solar masses.

Disconcertingly, however, Haswell and Shafter discovered a problem with what they had hoped would be a fairly straightforward case. They were able to discern, to no one's surprise, that when the orange dwarf approaches us the accretion disk recedes from us; when the orange dwarf recedes from us the accretion disk approaches us. The problem was that the black hole and its accretion disk seemed to be running more than an hour behind

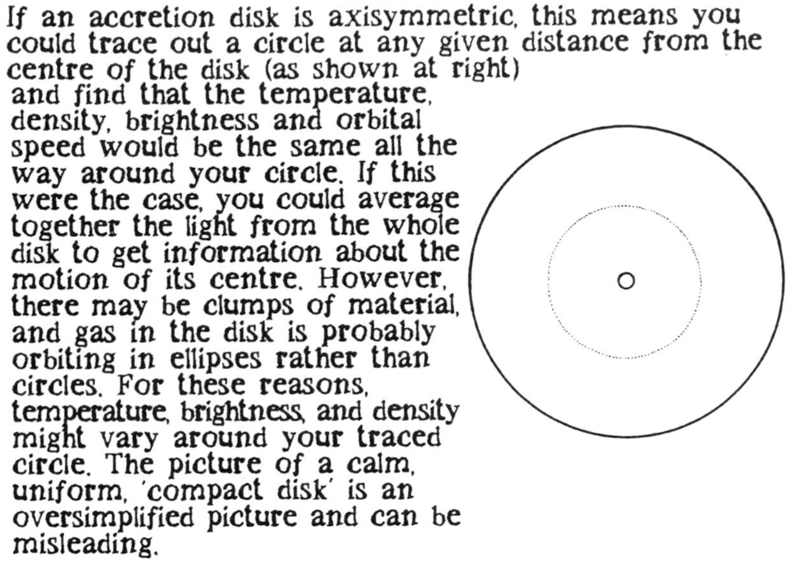

If an accretion disk is axisymmetric, this means you could trace out a circle at any given distance from the centre of the disk (as shown at right) and find that the temperature, density, brightness and orbital speed would be the same all the way around your circle. If this were the case, you could average together the light from the whole disk to get information about the motion of its centre. However, there may be clumps of material, and gas in the disk is probably orbiting in ellipses rather than circles. For these reasons, temperature, brightness, and density might vary around your traced circle. The picture of a calm, uniform, 'compact disk' is an oversimplified picture and can be misleading.

Figure 7.6.

schedule in orbit. If this is Wheeler's ballroom, the gentleman in this pair is behind the beat. However, the black hole and its surrounding disk cannot actually be lagging behind. Haswell and Shafter attribute the discrepancy to the fact that the emission of light from the disk is not symmetrical, meaning that you can't depend on the overall average of the light coming from the disk to tell you precisely where the black hole is (Figure 7.6). As Haswell explains, if you took the average location of all the residents of Dallas and assumed they are distributed symmetrically around the downtown area, you could come pretty close to pinpointing the location of downtown Dallas. However, this wouldn't work for Chicago, where the population is not distributed symmetrically around the centre of town. No one lives in the lake. Haswell and Shafter had been working on the assumption that A0620-00's accretion disk is Dallas, but, if their present suspicions are correct, it might be a little more like Chicago.

Haswell and Shafter had concluded that the black hole must be eleven times more massive than the orange dwarf, and so the orange dwarf's distance from the centre of mass must be eleven times greater than the distance of the black hole from the centre of mass. That result has now been superseded by a study of the spectra of light from A0620-00's accretion disk done in 1994 by T.R. Marsh, E.L.Robinson, and J.H. Wood of Oxford University, who put the mass ratio at fifteen-to-one, rather than eleven-to-one. The discrepancy is almost certainly due to the problem Haswell and Shafter suspect: a lack of symmetry in the disk's emission.

The study of A0620-00 has also had something to offer on the problem of judging whether we are viewing a system edge-on, and, if not, how far the system is tipped. Tim Naylor and Tariz Shahbaz of the University of Keele and Philip Charles of Oxford University and the Royal Greenwich Observatory had that dilemma in mind in 1993 when they took another approach to measuring the mass of A0620-00. McClintock and his colleagues had noticed earlier that the brightness of the orange dwarf varies, which led them to suspect that the orange dwarf is not a sphere. Tidal forces (see Chapter 4) may be distorting it into a teardrop shape. In fact, they suspect this is also happening in other binary systems that may harbour black holes. There has been speculation that studying how great the variations in brightness are may tell us something about the mass of its compact companion.

As the star orbits, the pointed side of the teardrop always points toward the compact companion, and the visible star thus presents different profiles to us on earth (Figure 7.7). For instance, if we are viewing the system edge-on, then twice during each revolution we see the teardrop shape from the side and the star's apparent brightness is at its maximum. Twice during each revolution, we see it end-on and its apparent brightness reaches a minimum. Assuming that this explanation for the variations in brightness is correct, Naylor, Shahbaz and Charles suggest that we might find out how close or far we are from viewing the system edge-on by noticing just how much the visible star appears distorted. They reason that a smaller variation in the brightness of the light signifies less range of distortion, and the less range of distortion there is, the farther from edge-on we are viewing the system. Naylor announced in late March 1993 that A0620-00 had weighed in at 7 to 22 solar masses.

Others think this approach has led to an overestimation of how

The orange dwarf companion of A0620–00 varies in brightness, which leads astronomers to suspect that the orange dwarf is not a sphere but is distorted into a teardrop shape by tidal effects. There is similar variation in other binary systems that may harbour black holes. The idea is as follows: the pointed side of the teardrop shaped star always points toward the compact star, and so the teardrop shaped star presents different profiles to us on earth as it orbits. Twice during each orbit we see it from the side and its apparent brightness is at its maximum. Twice during each orbit we see it nearest to end-on and its apparent brightness is at a minimum. Hence, the greater the variation in brightness, the closer to edge-on we are viewing the system.

In a system viewed edge-on, or very near to that, we find the stars eclipsing one another.

In a system viewed relatively near to edge-on, the variations in brightness are large.

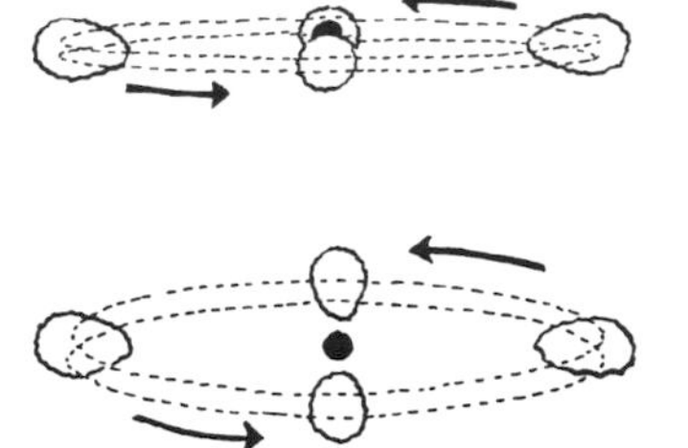

In a system viewed very far from edge on, we can expect very little variation in brightness.

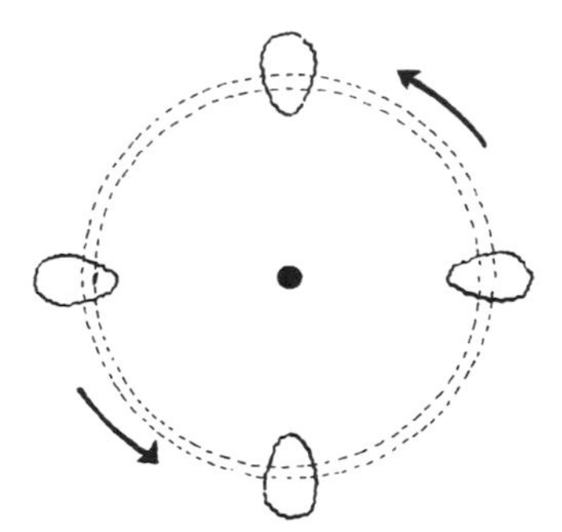

How great the variation in brightness is may help us to find out how far from edge-on we are viewing a system. The greater the variation in brightness, the closer to edge-on our view is.

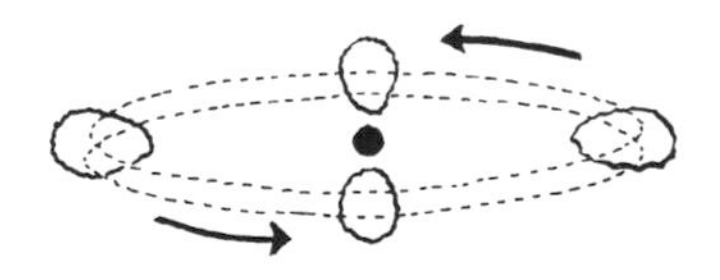

Figure 7.7. One possible method for determining how much a binary system is ‘tipped’, that is, how close to edge-on we are viewing the system.

far our viewing angle is from edge-on and how large the mass of the black hole is. Though A0620-00 is not an eclipsing binary (a binary in which the partners take turns eclipsing one another), Haswell and others have observed what they think may be 'grazing eclipses' in the optical light curves, which lead them to believe that the viewing angle must be closer to edge-on than Naylor, Shahbaz and Charles estimate.

Clearly there is still disagreement about the mass of A0620-00. However, no one at present is suggesting that the mass is small enough to allow this to be a neutron star. A0620-00 is a black hole.

V404 Cygni

Another X-ray source in the constellation Cygnus, GS2023+338, previously known as Nova V404 Cygni and now usually dubbed simply V404 Cygni, drew attention to itself in 1989 with an intense burst of X-rays that was detected by the Japanese X-ray satellite Ginga. In the satellite observation, Philip Charles noticed X-rays in the high-energy range that he suspected were part of the pattern we've mentioned as a possible black hole fingerprint. He and his colleagues Jorge Casares and Tim Naylor were encouraged enough to try to measure this candidate's mass by continuing to monitor the movements of its partner. They found the partner orbiting so rapidly that, even if it weighs almost nothing at all, it has to be in the gravitational clutches of a compact companion star with a mass of not less than six solar masses. Charles believes it would be realistic to estimate the mass at between 8 and 15 solar masses. This is not a borderline case.

More recently, a team led by D. Sanwal and Edward L. Robinson of the University of Texas deduced an upper limit of about 12 solar masses for V404 Cyg. There is also the interesting possibility that V404 Cyg is a system of two stars and a black hole – a triple system.

Looking toward the future

A new generation of large telescopes should allow us to get better data and clear up some of the current disagreements about the masses of compact companion stars. The Hubble Space Telescope is primed to spend twenty-five orbits – that's a day and a half – following the next intense outburst of X-rays from a promising candidate, focused on the source of the outburst.

Watch also for news of these other strong black hole candidates in binary systems:

* LMC X-3 in the Large Magellanic Cloud, with a minimum mass of 3 solar masses and probable mass between 4 and 11 solar masses.
* X-ray Nova Sco 1994, the only strong black hole candidate that we view edge-on. Its mass is 4 or 5 solar masses.
* X-ray Nova Oph 1977, a candidate whose minimum mass is 3 solar masses.

8

Hearts of darkness

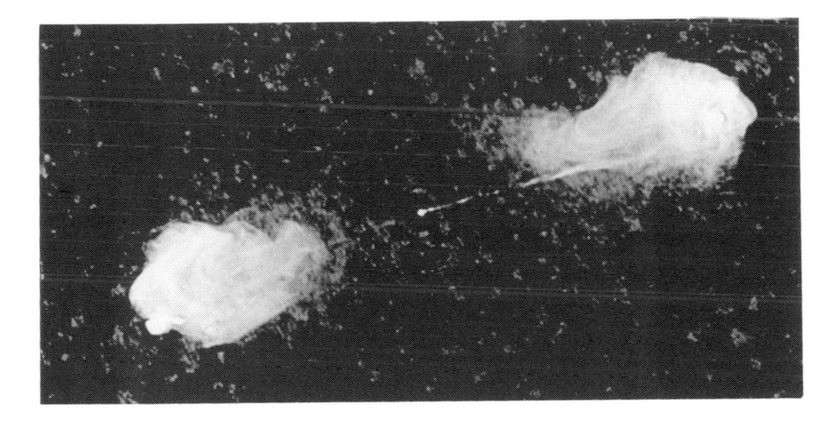

Cygnus A, a double radio source. From its core, a tiny dot at the centre of this image, narrow jets of hot gas extend over vast stretches of space, ending in puffy, turbulent lobes. This radio-telescope image is from the US Very Large Array (VLA). Through an optical telescope we would see a giant elliptical galaxy approximately the size of the region between the two lobes, centred at the dot. (National Radio Astronomy Observatory/AUI.)

The public doesn't understand what a human enterprise science is. It's like following Columbo on the chase. That's where the real excitement is.

Tod Lauer

The lights at the edge of spacetime

Only comparatively recently have we realized that when we gaze at the stars we're looking at the past. Light and other radiation don't reach us instantaneously across space. They travel at approximately 300,000 kilometres (186,000 miles) per second – not a speed to be scoffed at – but the distances involved are so enormous that it takes billions of years for images of the most remote objects to flash across the universe to our telescopes, computer screens, and photographic plates. When such ancient light left its source, there was no human pondering the night sky. There is no way these images could get here quicker. Building more powerful telescopes doesn't speed them up. We are like settlers in the old West waiting for the stagecoach to arrive, having to content ourselves with old news.

What might seem a frustrating delay in information and a severe limitation on our ability to find out what is happening in the universe turns out to be an advantage. We're able to study the past. The further 'out' we look, the further back we peer into the history of the universe.

Some of the objects most distant from us in both space and time are quasars. None is near us. They are all incredibly far away and long ago, shining like beacons to us from the dawn of spacetime. The most remote of them we observe nearly at the limit of the observable universe, which means we can only know them as they existed when the universe was a very small percentage of its present age. There is no way to find out whether these strange objects still exist today. If they do, they may by now have evolved into something quite different from the way they appear to us. How do they appear to us? Quasars observed from earth look like faint stars but also in some ways resemble gaseous nebulae. They are very small by cosmic standards but extremely bright, varying in brightness, with an enormous redshift.

When astronomers Alan Sandage, Thomas Matthews and Maarten Schmidt of the California Institute of Technology first studied these objects in the early 1960s, it seemed that existing physics might not be able to explain them. Three possible ways of accounting for their huge redshift all presented problems:

1. The redshift might be caused by a gravitational field. (Recall our discussion of gravitational redshift in Chapter 4.) If so, these objects would have to be so massive and so near to us that they would disturb the orbits of the planets in our solar system. Clearly that wasn't happening.

2. The objects might be stars in our own galaxy, ejected from somewhere with a force powerful enough to cause them to move away from us at a speed necessary for the observed redshift. (Also see Chapter 4.) The objects' spectra made this seem highly unlikely.

3. As the universe expands, galaxies and clusters of galaxies recede from us and from one another with the expansion. The more distant they are, the faster they move away from us. To cause such a redshift as astronomers were measuring for these mysterious objects, they had to be receding at as high as 37% the speed of light. For such a redshift to be caused by the expansion of the universe, they would have to be a vast distance from us. How then could we be observing them as we do? For instance, one of the objects under study was 3C273. If the reason for its redshift was the expansion of the universe, the redshift indicated that it was some 2 billion light years away. Yet 3C273 appeared as early as 1895 in photographs taken with optical telescopes of modest size. In order for 3C273 to be this readily detectable and as bright as it appears, and also as far away as its redshift indicates, it has to be radiating 100 times more power than the most luminous galaxies. How large *are* these things?

The variations in brightness provided a way of finding out. No source of light can flicker any faster than radiation can cross it. If it did, the next flicker would begin before the last one ended and the flicker would appear blurred. We wouldn't consider it a flicker at all. Keeping that in mind, remember that radiation cannot cross anything at a speed faster than the speed of light. Astronomers found that the light output of 3C273 changed substantially in periods as short as one month. That means most of the light from 3C273 must come from a region no larger than the distance light travels in a month. 3C273's size could not account for its brightness.

Using this line of reasoning, researchers found that quasars are tiny by cosmic standards, but many turn out to have even greater redshifts than 3C273. At these enormous distances, for us to observe such small objects as we do, they must be by far the brightest things in the universe, as bright as dozens or even hundreds of galaxies combined. Yet light from galaxies comes from regions of perhaps 100,000 light years across; while we know that the light from 3C273 comes from a region only one light month across.

For a time quasars were one of the foremost mysteries in astrophysics: how could anything so tiny radiate such energy, at all wavelengths? They must be incredibly violent, and for them to have a nebula-like appearance there must be huge regions of gas being ionized, superheated, and knocked about. It seemed they might indeed require a whole new physics to explain them. However, we now know that the source of power great enough to account for the energy a quasar emits could be a black hole of hundreds of millions or even billions of solar masses. There is increasing evidence that this explanation is the right one.

Because of their distance from us, quasars have been difficult to observe and study. Much of what we know about them we have learned by studying their close cousins, active galaxies.

Those who study galaxies separate them into two broad categories: active and normal. Our own Milky Way is a normal galaxy. Though there is dramatic activity at its core, we observe no great variations in brightness. By contrast, active galaxies flicker in the manner we have described quasars doing, varying in brightness over days, months and years. Their cores undergo violent, energetic upheavals; gas and stars there appear to be whirling rapidly.

Though quasars don't resemble active galaxies when viewed through optical telescopes, radio telescope pictures reveal a great similarity. At the core of an active galaxy or quasar is a faint, tiny source of radio waves. From this, narrow jets emerge, extending over vast stretches of space. The jets end in huge, puffy, turbulent lobes. It all looks a bit like a great, cosmic, double-ended chicken drumstick (Chapter 8 frontispiece).

Experts studying such phenomena concluded that something at the heart of a quasar or active galaxy shoots out hot gas in the form of the two narrow jets. Far away from the centre, the jets of gas slow down as they meet somewhat cooler gas, the 'intergalactic medium', which is much thinner than the most nearly perfect vacuum we can create in a laboratory. The hot gas bunches up and forms the lobes. We can think of this process as something like a teapot billowing steam into a cold kitchen. At first the steam forms a narrow pillar. Further away from the spout, the steam meets cooler air, slows down, and forms a turbulent lobe before disappearing. The steam looks as though it were meeting an invisible barrier (Figure 8.1).

Clearly, the engines that are accomplishing all this at the hearts of quasars and active galaxies are not very large: the funnelling starts off in an area not much bigger than our solar system. We now also know that these sources are massive, sometimes several billion solar masses. They are capable of channelling hot gas and delivering it over a time span of millions of years, at a good fraction of the speed of light, to distances over a billion times further away than the size of the 'engine' itself.

A black hole in residence?

There was speculation in the early 1960s that the collapse of a supermassive star or a large group of stars could be responsible for the amount of energy a quasar produces. The notion that the engine might actually be a gigantic black hole *resulting* from such a collapse came in 1964 from Edwin Salpeter at Cornell University and Zel'dovich in Moscow. In 1975, Jim Bardeen and Jacobus Petterson of Yale suggested a partial solution to the puzzle of how such a black hole engine might operate.

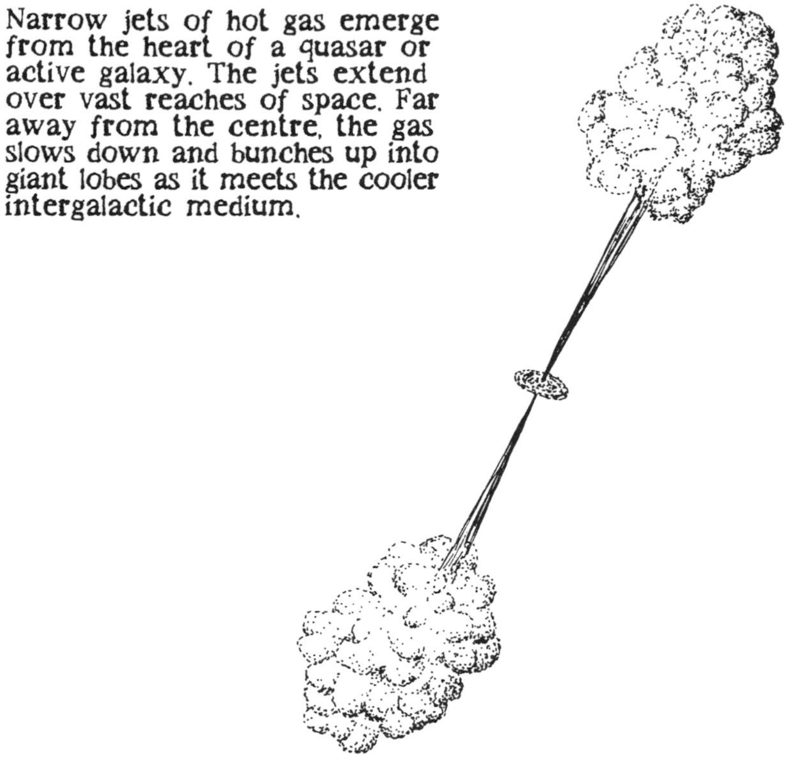

Figure 8.1.

Recall our discussion of 'hazards of rotation' in Chapter 4. A rapidly spinning black hole acts like a gyroscope. The direction it rotates never changes and creates a swirl of space nearby which also remains always oriented in the same direction. It is this swirl of space that causes the inmost part of an accretion disk to spin around a black hole's equator – in the hole's equatorial plane. Further from the black hole the capture of new gas may cause the accretion disk to tip, but that doesn't affect the disk's orientation nearer the black hole. The black hole's gyroscopic action keeps that inner part of the accretion disk always spinning in the hole's equatorial plane (Figure 8.2). We are going to see that this is important when it comes to setting the direction of the jets.

While the black hole thus orients the accretion disk, the accretion

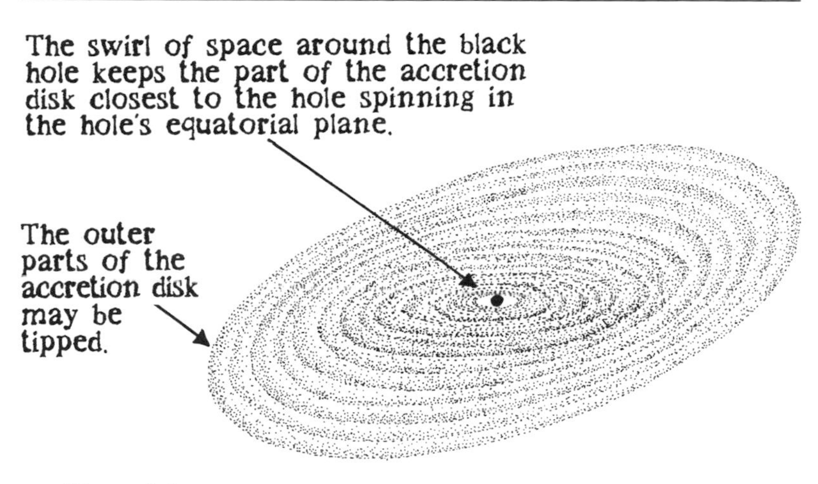

Figure 8.2.

disk also has an affect on the black hole. We saw in a previous chapter that it is possible for infalling particles to 'spin up' a black hole. Gas plunging into the hole from the accretion disk gradually makes it spin faster and faster until it reaches a point where centrifugal forces prevent any further increase in speed.

By what process, then, could this black hole 'engine' produce the jets we observe? In the mid-1970s, Roger Blandford, Martin Rees, Lynden-Bell and Roman Znajek at Cambridge came up with not one but four suggestions, and a combination of these methods may be in operation in quasars and in the cores of active galaxies (Figure 8.3).

1. Wind off the inner part of the accretion disk blows a bubble above and a bubble below the disk in a surrounding cloud of cooler gas that isn't part of the disk itself. Hot gas from inside those bubbles then punches openings through the cooler gas cloud, and hot gas shoots out from these openings. The direction the hot gas takes, as it joins in a jet the way water does coming out of a faucet, is along the spin axis of the hole, that is, the directions pointing straight 'up' from its 'north' pole and straight 'down' from its 'south' pole. The gas will thus shoot off in two oppositely pointed directions, forming the jets.
2. Pressure of the accretion disk's internal heat may puff the

(1) Wind coming off the inner part of the disk creates a bubble of hot gas in a surrounding cloud of cooler gas. The hot gas in the bubble punches orifices through the cooler cloud, probably along the spin axis of the black hole (the directions of its north and south poles). Hot gas shoots out, forming the jets.

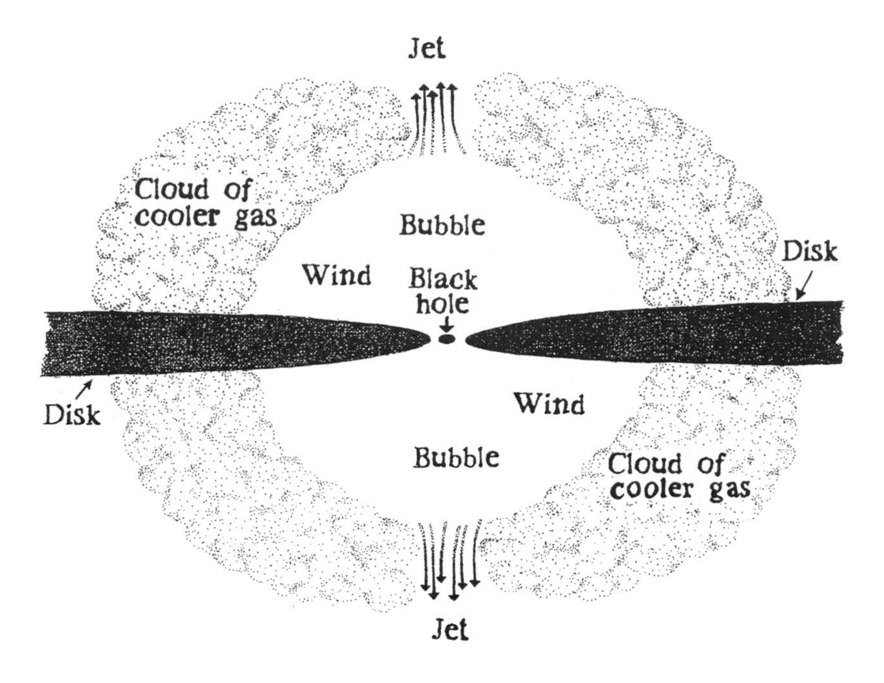

Figure 8.3. How could a black hole 'engine' produce the jets we observe? Here are four possible explanations. A combination of them may be in operation in quasars and in the cores of active galaxies.

disk up until it is more like a bagel with a very small centre hole than a compact disk. The upward and downward funnels at the centre of the bagel focus gas off the disk into two jets.

3. All gas is magnetized, and when it comes into an accretion disk it carries its magnetic fields along. Magnetic field lines anchored in the accretion disk and sticking out of it into the centre 'funnels' of the 'bagel' are forced by the disk's

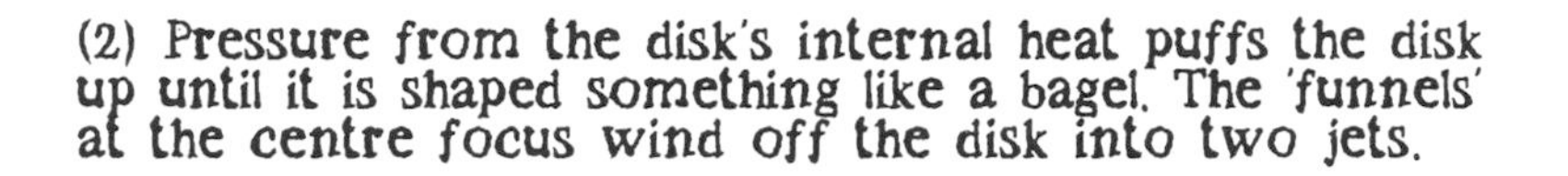

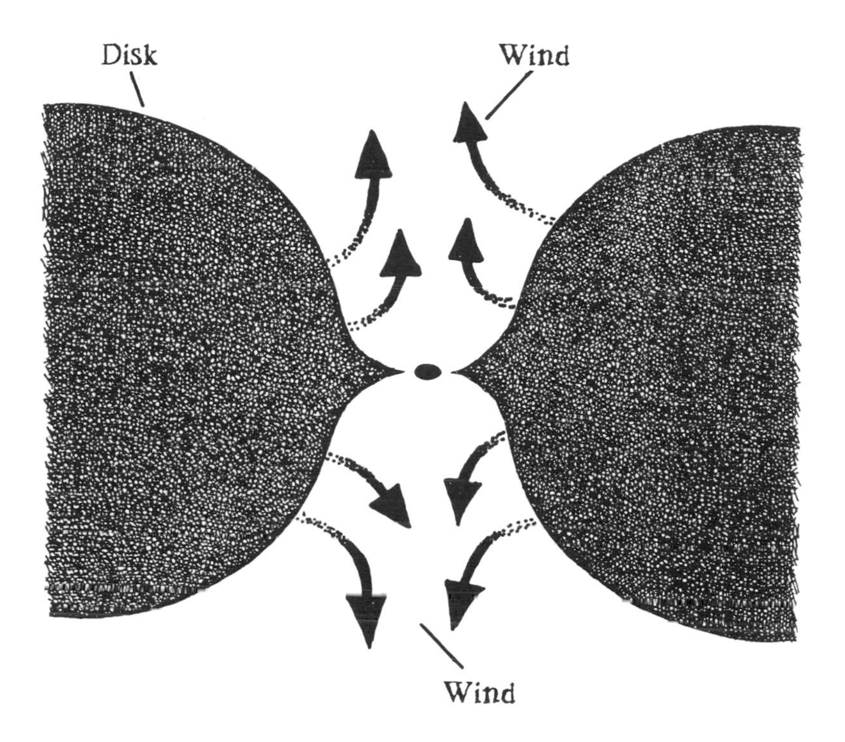

rotation to spin and form spirals leading out of both funnels. Hot gas slides along the field lines and, as the field lines spin, centrifugal forces throw the hot gas outward along them, forming the two jets.

4. This method is similar to number 3, except that now the field lines stick out of the black hole itself, not out of the disk. Again, all gas is magnetized, and when it eventually falls into the black hole, it carries its magnetic fields along. As it nears the black hole, a bit of gas slides down its magnetic field lines and through the event horizon, leaving the magnetic field lines sticking out of the event horizon. As the hole spins, it drags these field lines around with it. The field lines form spirals as they did in method 3.

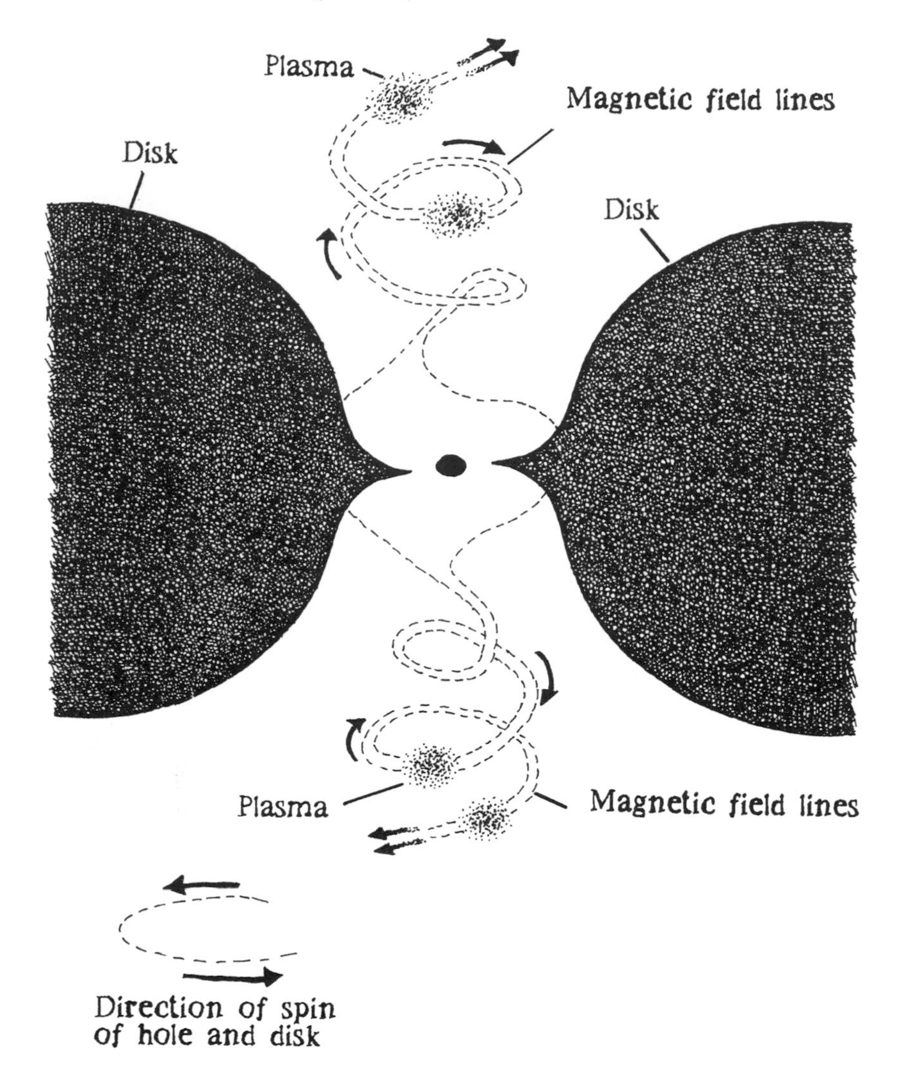
(3) The orbital motion of the disk forces magnetic field lines anchored in the disk and sticking out of it to spin, and the spinning field lines spiral outward from both 'funnels'. The field lines fling hot gas (plasma) upward and downward, forming two jets.
Plasma
Magnetic field lines
Disk
Disk
Plasma
Magnetic field lines
Direction of spin
of hole and disk

(4) Similar to (3), except that now the field lines stick out of the black hole itself, not out of the disk. When gas falls into a black hole it carries its magnetic fields with it. As it nears the black hole, a bit of gas slides down its magnetic field lines and through the event horizon, leaving the magnetic field lines sticking out of the event horizon. As the hole spins, it drags the field lines around, causing them to spiral and fling plasma outward as in (3).

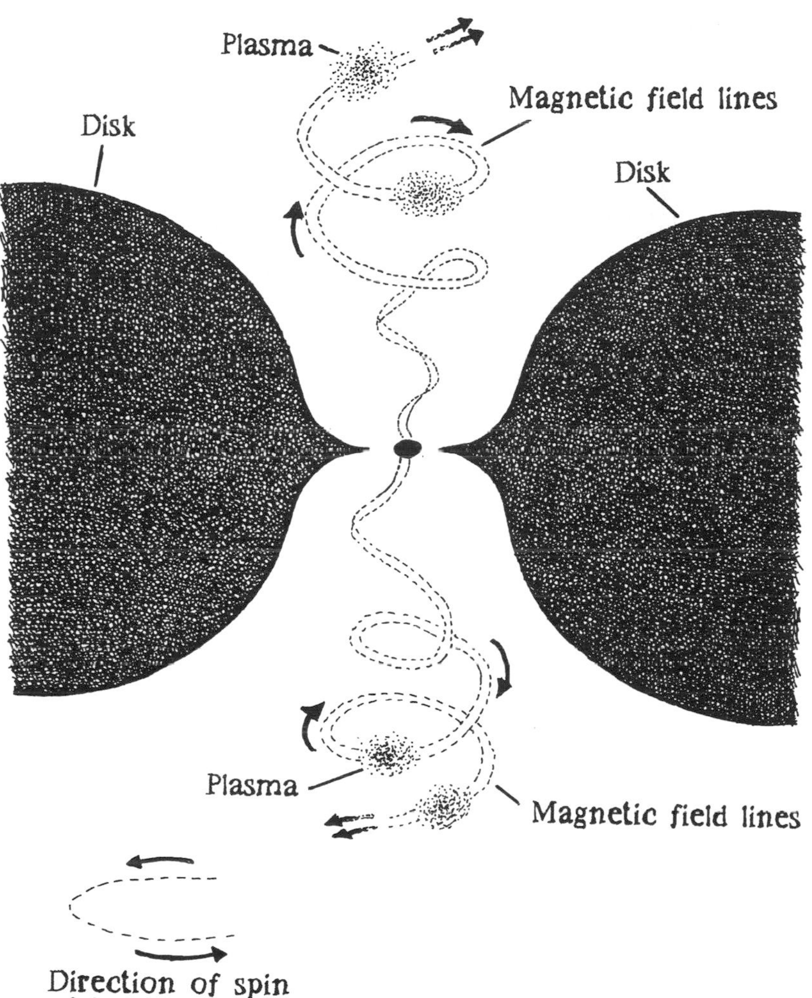

In all four cases, the spin axis of the hole dictates the direction of the jets.

An important question remains before we look at some recent observational discoveries having to do with quasars and the centres of galaxies. How could such a monster black hole form? If it were from the collapse of one star, that star would have to have been supermassive indeed. There are more likely possibilities.

Martin Rees argues that massive black holes probably developed routinely during the process in which enormous clouds of gas gathered into galaxies in the early universe. Other suggestions involve both stars and gas. It seems that not all denizens of Wheeler's ballroom limit themselves to sedate waltzes. When two stars pass close to one another their gravitational forces can cause them to swing around each other and, rather than linking them in a binary system, fling them off in directions which are different from the paths they were following previously. Frequently one will end up heading inward toward the galactic centre while the other heads outward. As these events are repeated, some stars are driven deeper and deeper into the galaxy's heart. At the same time, friction in interstellar gas sends much of that gas also down into the core. More and more stars and gas accumulate in the core of the galaxy, and the gravity of their combined mass becomes increasingly strong – eventually strong enough to implode and form a giant black hole. Yet another suggestion has it that individual massive stars among those crowded in the core collapse to form black holes. Then these holes collide with one another and with stars and gas, resulting in larger and larger holes until there is at last one single gigantic black hole. With any of these scenarios or all combined, it's plausible that many galaxies by now have black holes in their centres.

As to the origin of black holes in quasars, one idea that has been gaining ground is that quasars are born in collisions between large, gas-rich galaxies of similar masses. Each of the galaxies may already have a black hole at its centre, or the collision may create a black hole. We observe collisions between galaxies, and many more could have occurred in an early era when the universe was smaller and more crowded. Since a more accepted view is that quasars represent an extremely active period in the early development of many galaxies, this becomes a bit of a chicken and egg problem which will probably not be easy to resolve.

Almost sure to be a black hole . . .

By whatever method the monsters have come to be there, in the late 1970s and the 1980s it began to look more and more likely that they are, indeed, there. Evidence continued to accumulate that black holes exist in the cores of most quasars and active galaxies and possibly also in the cores of normal galaxies such as the Milky Way, Andromeda, and smaller galaxies.

In 1987, German astronomer Norbert Bartel and his associates studied an active galaxy called 3C84 at high enough resolution to observe a region at its centre less than a tenth the size of any studied previously. What they found looked like a sphere surrounded by an extended pancake-shaped ring. There was too much radio emission coming from such a small region for this to be a collection of many stars. Bartel thought it must be a single compact star more than a million times more massive than the sun, surrounded by an accretion disk. Such a star seemed almost sure to be a black hole.

Five years later, in the fall of 1992, there was tremendous excitement when the Hubble Space Telescope identified NGC4261, an elliptical galaxy in the cluster known as Virgo, about 45 million light years from earth, as harbouring the most likely black hole suspect to date in a galactic centre (Figure 8.4). In orbit above the earth's atmosphere, even with faulty optics before its repair, the Hubble was able to reveal much more about the universe than ground-based optical telescopes, and Bruce Margon of the University of Washington said about the discovery in NGC4261, 'We now see something that is almost too good to be true.' Astronomer Walter Jaffe of the University of Leiden in the Netherlands announced, '[This] is the best look we have ever had at the workings of the nuclear engine at the centre of an active galaxy. We haven't seen the black hole itself, but we are seeing closer to a black hole than we have ever seen before.'

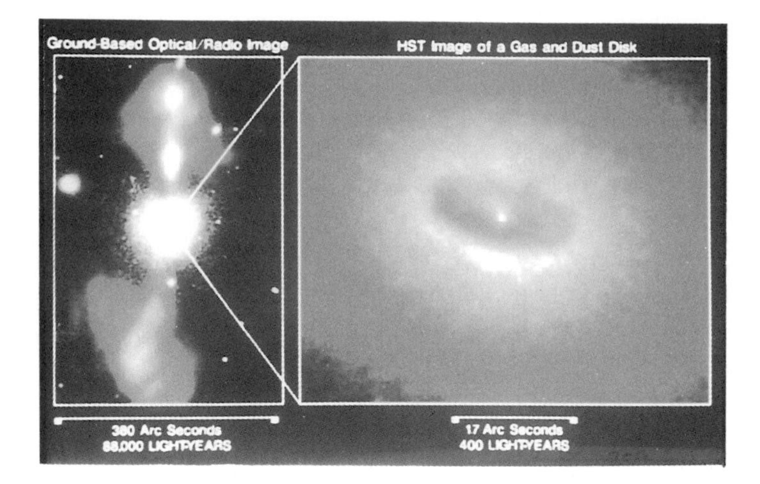

Figure 8.4. Galaxy NGC4261. On the left is a composite optical/radio view: photographed in visible light, the galaxy appears as a fuzzy disk of hundreds of billions of stars (centre of picture), while a radio image shows a pair of jets emanating from the nucleus and spanning a distance of 88,000 light years. On the right is the Hubble Space Telescope image, taken at visible wavelengths in 1992, before the repair of the telescope. It shows a giant disk of gas and dust fuelling a possible black hole at the core of the galaxy. (Left: NRAO-CalTech. Right: Holland Ford/HST/NASA.)

Jaffe and his American colleague Holland Ford of Johns Hopkins University and the Space Science Institute had decided to train the telescope on this particular galaxy becausc it is one of the brightest in its cluster, containing about 100 billion stars. What they discovered at the centre is a swirling, dark, dusty disk which is about 300 light years in diameter. Fortunately, the disk is tilted in such a way as to provide a clear view of the galaxy's hub.

The outer fringe of the disk is rotating at about 80 kilometres per second, while the inner region of the disk possibly is rotating at approximately 5000 km per second. 'This is the first case where we can follow the disk's gas in an orderly way down to the immediate environment of the black hole,' said Ford.

The Hubble telescope examines visible and ultraviolet light. NGC4261 had also been observed with radio telescopes, and these observations showed that the core is shooting out jets of energy extending across at least 88,000 light years of space. It seems the material in the disk reaches a temperature of tens of millions of degrees as the speed of its motion increases and as it is compressed by gravity near the centre. Before falling into the black hole, some of the hot gas squirts out and forms the jets in a process like those described earlier.

What other evidence, besides that coming from the 1992 Hubble observations and radio telescopes, would be needed in order to say absolutely that NGC4261 is a black hole? More precise measurements of how rapidly the different parts of the disk are rotating, how the stars in the galaxy, especially those near the centre, are moving, and particularly the motion of the gas within a few dozen light years of the suspected black hole. In 1992 Margon, looking forward to the repairs to the Hubble telescope and measurements that would be taken with an instrument on board called a 'Faint Objects Spectrograph', said, 'What is going to convince most people is to actually know the speed of the stars in the very centre of this galaxy. Then, using that measurement and Newton's laws of gravity, we can directly weigh the central object. If we discover that it is extremely massive . . . and extremely compact, then there will probably be no alternative to a black hole.'

That was how things stood when the Hubble telescope, wearing corrective lenses installed during a highly publicized repair mission in December, 1993, began with clearer vision to scan the skies in the winter of 1994.

The duck quacks

More than 50 million light years from earth in the constellation Virgo there is an elliptical galaxy known as M87. M87 is made up

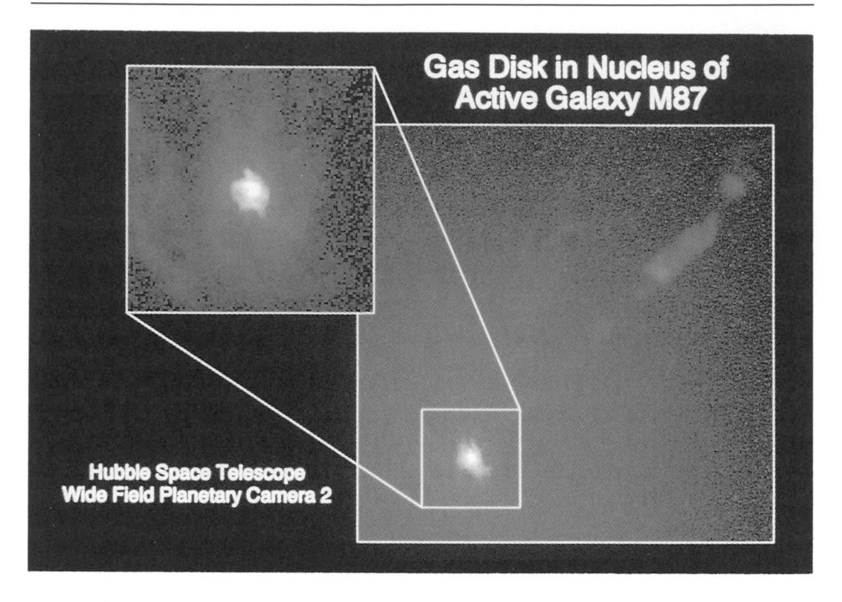

Figure 8.5. Galaxy M87. Left: The orbiting disk of hot gas at its heart, photographed by the repaired Hubble Space Telescope in 1994. Right: Previously known jet emanating from the galaxy's centre. (Ford, Harms *et al.*/HST/NASA.)

of close to a trillion stars and is a powerful emitter of radio waves. It has been of particular fascination to astronomers since early in this century when they detected a jet of light shooting out of its core (Figure 8.5), and it became a prime suspect for harbouring a massive black hole when theorists suggested in the 1970s that such jets of hot gas should be considered important black hole clues.

In 1978, Wallace Sargent of the California Institute of Technology and co-workers analysed spectra from the core of M87, attempting to find out the velocities of the stars there. Studying star velocities in an elliptical galaxy presents a problem because the stars move in many directions, not in the more orderly way they do in a spiral galaxy such as our own. Nevertheless, Sargent and his colleagues felt confident that the velocities they found were too great for the visible matter alone to account for them – a strong argument for the presence of a black hole. The velocities were consistent with the presence of a 5-billion-solar-mass object at the core.

Meanwhile Peter J. Young, also of Cal Tech, and colleagues studied M87's light profile. The gravitational tug of a black hole at the core of a galaxy could be expected to pack stars so densely that the intensity of starlight would rise dramatically close to the centre. Young said that the steep rise in intensity of light he found toward the centre of M87 suggested a 3-billion-solar-mass black hole. Further analysis showed that although there might be other explanations for Sargent's and Young's findings besides a black hole, the other models ought to have evolved to produce a bar-like structure, which is not present in M87.

In 1992, the Hubble telescope entered the M87 story. Tod Lauer and Sandra Faber of the University of California, Santa Cruz, after making corrections to adjust for the telescope's faulty optics, confirmed that the brightness of starlight did indeed increase dramatically near the centre. Stars seemed to be crowded there so tightly that their density exceeded 300 times that usually found in giant elliptical galaxies. This was strong circumstantial evidence, but still a black hole was not the only possible explanation. Theorists could construct models of galaxies with unusual distributions of stars that would produce a similar bright pinpoint with no black hole present. Lauer himself cautioned that some other feature such as a massive flow of gas into M87's centre might draw in a dense collection of stars. 'It looks like a duck,' he quipped, 'but we haven't heard it quack yet.' To prove the actual existence of a black hole in residence it was necessary to observe the starlight and swirling gases very close to the galactic nucleus in sufficient detail so that their orbital velocities could be measured precisely. If there is a massive black hole there, matter should accelerate to near light speed as it falls in toward the core. Prior to December 1993, the Hubble telescope with its faulty optics could not produce sufficiently detailed pictures.

In March 1994, a team led by Holland Ford and Richard Harms pointed the newly repaired Hubble at the centre of M87, and in May they announced that they had been able to see clearly into the heart of that galaxy (Figure 8.5). The camera had penetrated four times closer to the centre than any ever had before. There was no need to sort through closely packed stars and their overlapping light. Ford said he was astonished to see through to a large, well-ordered, pancake-shaped disk of hydrogen gas, stretched across 500 light years, spinning around and being consumed by something at the centre. The disk reached within 60 light years of the galactic nucleus. The telescope revealed such sharp details that it was possible to 'weigh' the invisible object at the centre quite easily. In order to find the velocities crucial to this measurement, the researchers used the Hubble's Faint Object Spectrograph. This apparatus split the light of the whirling disk into different wavelengths as it was emitted from opposite sides of the disk (the side on which the gas approaches us and the side on which it is moving away), and experts studied the red and blue shifts. They found that at its inner edge, the disk is rotating at about 1.2 million miles per hour (750 km per second). What central mass would be required to keep the disk spinning so rapidly without flying apart? Calculations indicate that the object must have a mass of 2 billion to 3 billion solar masses. 'This is the mother of all black holes,' quipped Harms. That mass is confined in a region no larger than our solar system. Observations also show that the jet is at a right angle to the disk, as expected.

Are we *sure* it's a black hole this time?

Edwin Salpeter, who in 1964 with Zel'dovich had predicted discoveries such as this, commented about the new Hubble data: 'A black hole is now the least crazy model for what we're seeing. It's good to finally win the bet.' Ford told a news conference, 'This is conclusive evidence of a supermassive black hole.' Nobody was arguing with him. Even Daniel Weedman, NASA's director of astrophysics, who had been sceptical of previous evidence regarding black holes, was convinced. 'I do believe,' he said, 'there is a black hole there.'

The visible light at the centre of the Hubble image is not starlight. Just outside the event horizon, superheated, ionized gas clashes with magnetic fields generated as the black hole and the space around it spin in tandem. Electrons in the gas accelerating to near light speed through the magnetic fields emit a light known as synchrotron radiation. It is this radiation that we see as white-hot light at the centre.

The hunt continues for black holes in the hearts of galaxies and quasars. Even as he was announcing the dramatic discovery in the heart of M87, Ford said he was looking forward to training the repaired telescope on NGC4261, the galaxy he and his colleagues had studied with such intriguing results in 1992 before the Hubble telescope's corrective lenses were installed, and on the Andromeda galaxy (M31), close to our own galaxy and much like it. The Andromeda galaxy and NGC3115, another nearby galaxy, show evidence of rapidly orbiting stars at their cores, with speeds increasing dramatically toward the very centre. The mass of visible stars is not sufficient to explain these velocities in either galaxy. In 1995, Ford and Yichuan C. Pei of Johns Hopkins University announced that the Hubble telescope had found asymmetries at the core of the Andromeda galaxy that suggest its nucleus may be a binary system consisting of a black hole and a bright companion. They estimate the mass of the black hole to be about 6 million solar masses.

Also relatively near us is M32, a tiny satellite galaxy of the Andromeda galaxy. Since the 1980s, ground-based telescopes have shown that the orbital velocities of stars rise steeply toward its centre as if they are circling a massive object. Lauer studied it with the Hubble in 1992 and found that the millions of stars at its centre become extremely concentrated toward the nucleus of the galaxy. However, with the Hubble's faulty optics it was possible to resolve only the beginning of the rise in star density, and impossible to find out whether the density levels off further into the core. M32 has no central jets and its nucleus doesn't radiate intensely at any wavelength, which has led Lauer to suggest that smaller black holes such as this one may consume surrounding gas less frequently and spew out less energy – or spew it out in periodic bursts, not continually. As we'll see in Chapter 9, there is a possibility this is what happens with a black hole in the core of our own galaxy. Lauer suggests that in M32 the high density of stars stems from a relatively small black hole with a mass of about 3 million solar masses.

And then there is a curiosity known as M51, 20 million light

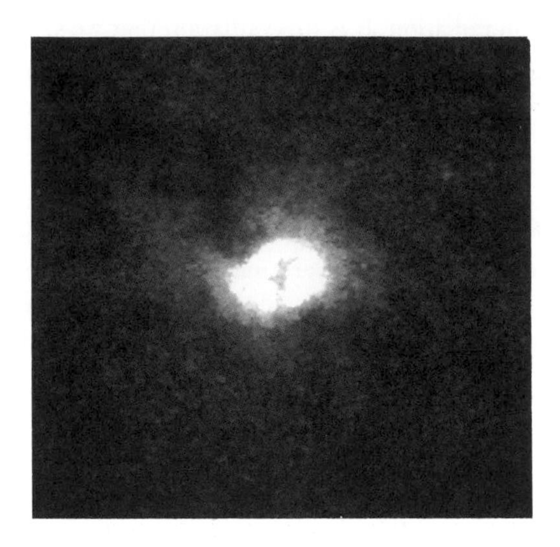

Figure 8.6. Galaxy M51, photographed by the Hubble telescope in 1991. The fatter arm of the 'cross' may be an edge-on view of a rotating doughnut-shaped ring of cold gas and dust around a central black hole (Holland Ford/HST.)

years away, which looks a little like a hot cross bun shown from the top in negative (Figure 8.6). When the Hubble telescope photographed it in 1991, it was dubbed 'X marks the spot' and astronomer Ed Weiler quipped: 'God wanted us to find a black hole, so he put a big X on it.' Instead of jets, M51 has what look something like a pair of cone-shaped searchlights streaming out from the centre in opposite directions. Holland Ford proposed in 1991 that the fatter arm of the 'cross' is an edge-on view of a rotating doughnut-shaped ring of cold gas and dust around a central black hole. Ford's team plans to use the Hubble Faint Object Spectrograph to try to find out more about emissions near that part of the X.

NGC4258 is also a close neighbour at a distance of 21 million light years, bright enough to have been catalogued as early as the eighteenth century. Findings from ground-based telescopes reported in January 1995 indicate a slightly warped disk of rotating gas near the centre of the galaxy. The observed rotation velocities imply the presence of a mass density at least 40 times that of any previously observed black hole candidate.

Clues in the gamma-ray spectrum

Interesting findings in remote regions of the universe have come from another orbiting observatory, the Compton Gamma Ray Observatory (GRO or Compton for short). As its name indicates, the Compton's telescopes scan the skies for sources of gamma radiation. Gamma rays are the most energetic form of electromagnetic radiation, but they cannot penetrate the earth's atmosphere (see the spectrum, Figure 3.1). Soon after the Compton's launch from the space shuttle Atlantis in 1991, R.C. Hartman and colleagues at the NASA/Goddard Space Flight Center discovered two 'blazars' – thought to be quasars aligned so that a jet points directly toward the earth. One of these, 3C279, about 4.6 billion light years away is, in spite of its enormous distance, one of the brightest sources in the high-energy gamma-ray sky. The brightness of 3C279 varies over a period of only a few days. You'll recall from our discussion of the way quasars and active galactic nuclei 'flicker', that such rapid variation means the size of the region where the radiation originates is very small. In the case of 3C279, it seems that a physical change can travel across the source region, causing a significant change in its gamma-ray emission, in a few days. The source region can probably be no larger than a few light-days across – only several times the diameter of Pluto's orbit around the sun.

The Compton has since found many more gamma-ray-emitting quasars and active galactic nuclei at distances ranging from 400 million to 9 billion light years away. The most remote are observed nearly at the limit of the observable universe. Why are these such strong gamma-ray sources? The theory is that low-energy photons – such as visible light or ultraviolet rays generated in the disk around the black hole – bounce off rapidly moving electrons in the jet, gaining enough energy to become gamma rays. The result would be a brilliant, focused beam of gamma rays, and the blazar would appear especially bright if the jet focusing the beam were pointing directly toward the earth.

9
The search goes on

View in the direction of the constellation Sagittarius. The box indicates the centre of the Milky Way galaxy, which we cannot see with our eyes or with optical telescopes because of intervening interstellar dust and debris. (Harvard College Observatory.)

The joy and fun of understanding the universe we bequeath to our grandchildren – and to their grandchildren. With over 90 per cent of the matter in the universe still to play with, even the sky will not be the limit.

Vera Rubin

Is there a monster in our basement?

When we look toward the constellation Sagittarius in the night sky, we are looking toward the heart of our Milky Way galaxy, about 28,000 light years away. Intervening clouds of interstellar dust prevent our having a spectacular optical view of billions upon billions of stars, tightly concentrated in and around a great bulge at the centre of the galactic spiral. Optical telescopes (telescopes that show us visible light) cannot penetrate those clouds. However, the galactic centre is a very compact source of radiation in other parts of the spectrum. With telescopes that detect these kinds of radiation we're able to probe the innermost few hundred light years.

Deep in the Milky Way's core there is a mysterious and remarkably powerful source of radio waves known as Sagittarius A* (pronounced 'Sagittarius A star'). Over the past two decades, studies of these radio waves and observations in other ranges of the spectrum have made an increasingly convincing case that Sgr A* is a massive black hole surrounded by an accretion disk – a scaled-down version of what we have discovered in the nuclei of active galaxies. After all, might we not be wondering, 'With black holes turning up in the centres of so many galaxies, why should we be different?' However, the case is not closed. Until recently there was one particularly troubling gap in the evidence. If Sgr A* is a black hole of the sort we suspect, we ought to be picking up some emission in the infrared part of the spectrum. Infrared radiation from Sagittarius A* eluded detection.

In the 1990s, this missing piece of the puzzle seems at last to be falling into place. In 1992, faint infrared emission coming from the very centre of the galaxy showed up in observations by Reinhard

Genzel and colleagues at the Max Planck Institute in Germany and in other studies by Laird M. Close and Donald W. McCarthy, Jr, of the University of Arizona, supporting the possibility that our galaxy's centre harbours a black hole of about 2 million solar masses. A year later, new infrared observations by Andreas Eckart and others at the Max Planck Institute, and again by Close and McCarthy, continued to yield persuasive evidence. Then, in 1994, Close and McCarthy produced even sharper and more definitive images, revealing what appears to be a fast-spinning accretion disk in the region of Sagittarius A*. The consensus now is that the object at the centre of the disk has a mass of about 1.3 million solar masses. From radio observations we know this massive object occupies a region no larger than our solar system, probably considerably smaller.

Are there any lingering doubts that our galaxy has a massive black hole in its core? Astronomers are still cautious.

First, although the infrared emissions come from the region of Sagittarius A*, there is a small possibility that they do not really come *from* Sagittarius A*. The galactic centre is a crowded place, and a chance alignment of Sagittarius A* with some unrelated source of infrared radiation can't be completely ruled out. It is, however, not very likely.

Second, there is the question whether Sagittarius A* is actually the centre of the galaxy. A good argument in its favour is that it stays essentially at rest when measured against the background stars in the centre of the galaxy. We would expect that to be true of the galactic centre but not of anything else in the galaxy, all of which ought to be orbiting around that centre.

Third, in 1994, the first high-sensitivity survey of the galactic core for 'hard X-rays' (X-rays from the more energetic end of the X-ray spectrum) did not find evidence of the hoped-for black hole. In our discussion of black holes in binary systems, we learned that one important clue in the search for black holes is X-ray radiation extending up into these energies. If there is a massive black hole in the centre of our galaxy, we can expect it to radiate hard X-rays and gamma rays. Earlier studies raised hopes by finding hard X-ray and gamma-ray sources near the galactic centre, but new observations with more sophisticated equipment argue that this radiation is from isolated black hole candidates in binary systems or drifting through dense molecular clouds – not from a central black hole. These findings don't rule out a black hole at the galactic centre, but, if one is there, it must be an unusually placid monster.

In the light of what we've observed with other strong black hole candidates, it wouldn't be too surprising to discover that the central massive black hole in our galaxy is dormant, with mass gathering in a surrounding accretion disk for a future outburst but not presently accreting into the black hole. You'll recall that black holes in binary systems – A0620-00 Monoceros and V404 Cyg, for example – have 'on' and 'off' states in their X-ray luminosity. In some cases the 'on' states are flare-ups that last only a short period every fifty years or so. If galactic nuclei also have such cycles, perhaps the 'active' nuclei we observe in other galaxies are different from the vastly more numerous 'normal' galactic nuclei only in that they are in the relatively rare 'on' state, while black holes at the core of 'normal' galaxies, such as Sagittarius A* in the Milky Way, are in the 'off' state at present.

Though the mystery of what lies at the heart of the Milky Way galaxy is not entirely solved, we are tantalizingly near. In 1997, the Hubble telescope will be given infrared capability. One of its first observations will be of the galactic centre. Within hours after those observations, Laird Close tells us, we will know whether or not there is a black hole in the centre of our galaxy.

So . . . we may have a monster in residence all right . . . but asleep. What if it wakes up? Should we worry? Black hole expert Kip Thorne (who won the bet with Hawking, you'll remember) has some comforting statistics to offer.

If it exists, the central black hole in our galaxy probably has a mass of no more than about 3 million solar masses. That mass would give it a circumference of about 50 million kilometres. The circumference of the earth's orbit around the sun is ten times that large. Clearly, though we speak of a 'gigantic black hole', this monster is tiny indeed in comparison with the size of the galaxy. Our earth isn't near the galactic centre. It's in a solar system that orbits the centre on an orbit with a circumference of 200,000 light years, about 30 billion times larger than the circumference of the black hole. Over a period of time equalling 100 million times the present age of the universe, the central hole might swallow a large fraction of the mass of the galaxy, but even if it did this, its circumference would grow only to about one light year, still 200,000 times smaller than the circumference of our orbit. Thorne points out that the galaxy will change during that same period in many ways that we can't presently predict, and other catastrophes will almost certainly befall the earth and its inhabitants. But his advice regarding the monster in the middle is . . . not to worry!

Tracking white-hot black holes

Stephen Hawking proposed in 1991 that there are black holes that didn't form from the collapse of stars or star clusters or from colliding galaxies (see Chapter 6). He dubbed them primordial black holes, because if they exist they are relics of the early universe when there were pressures that could compress even a small amount of matter – far less than the Chandrasekhar limit – enough to form a black hole.

Theories about the early universe suggest that the universe may have undergone what is known as a 'phase transition'. We observe a homely example of a phase transition when water boils or freezes. In a phase transition, a uniform medium develops irregularities. In the case of water, what begins uniformly as water develops bubbles of steam or lumps of ice. In the phase transition that occurred in the early universe, conditions could have been such that irregularities collapsed to form black holes.

If a primordial black hole started out somewhat smaller than the nucleus of an atom, with a mass no larger than that of a small mountain, it would have evaporated away completely by now. But if it were a little larger than that, about the size of an atomic nucleus, with the mass of a more substantial mountain (more than a few billion tons), it should still be evaporating strongly today. Given their size, the most promising way of searching for such black holes – arguably the only possible way – is to look for the evidence of their evaporation; that is, the Hawking radiation coming from them and especially their ultimate explosion. A large portion of the radiation at this late stage should be gamma rays – the most energetic variety of photons. (See electromagnetic spectrum, Figure 3.1.)

In the search for primordial black holes (or PBH's, as they are called for short), it is frustrating to discover that although we do find gamma rays travelling randomly through the universe we do not find an *excess* of gamma rays. In other words, we find them in amounts and with properties that can just as easily be explained in other ways besides the evaporation of primordial black holes. Is the theory incorrect in predicting the evaporation of black holes? Or are there simply not very many of them? Calculations by Hawking and Don Page, of the University of Alberta, Canada, show that, if tiny evaporating black holes exist, they are not particularly thick-off-the-ground. There can be at present no more than 300 of them on the average in each cubic light year of space. However, since gravity would draw PBH's toward other matter, they should be more common than average in and around galaxies, and less common than average in the areas between galaxies.

It would be a breakthrough to detect the gamma-ray emission signalling the farewell explosion of an individual PBH. Such a burst of radiation would have to be detected above the diffuse background of gamma rays from other sources, including evenly distributed evaporating PBH's. Would such an individual event stand out strongly enough for our present instruments to make the distinction? What would distinguish it conclusively from other gamma-ray bursts? These are questions that theorists and observational astronomers are presently trying to answer.

It's not possible to study gamma radiation from space directly from the ground because this radiation is absorbed by the earth's atmosphere (our crab problem, as mentioned in the Prologue). However, it is possible to study it indirectly by observing flashes of light in the night sky. When high-energy gamma rays hit the atoms in our atmosphere, they create pairs of electrons and positrons. These in turn hit other atoms, creating more pairs of electrons and positrons until there is an 'electron shower'. The result is a form of light known as Cerenkov radiation which shows up as flashes in the night sky. By observing the flashes simultaneously with telescopes at widely separated locations on the earth's surface, researchers are able to distinguish between these phenomena and flashes caused by lightning or reflections from satellites and orbiting debris.

In the 1970s, shortly after Hawking introduced the idea that primordial black hole explosions should occur and that an intense burst of gamma rays might be evidence of such an explosion, there were several experimental searches for this evidence, using ground-based telescopes. The most significant results were the statistical population limits mentioned above – an average of no more than 300 tiny evaporating black holes per cubic light year.

Now, in the 1990s, the Compton Gamma Ray Observatory, orbiting above the atmosphere, scans the cosmos for gamma radiation with unprecedented sensitivity. Of particular interest are gamma-ray flashes. These bright blips of gamma radiation are not uncommon, and we have known of their existence since the late 1960s. The Compton has discovered them to be everyday occurrences. However, just *what* they are remains one of the unsolved mysteries of modern astronomy, and the Compton findings have only added to the puzzle, not solved it.

Gamma-ray bursts are very short, usually lasting from about one hundredth of a second to 1000 seconds, and they differ from one another dramatically when it comes to brightness. However, the prediction is that a flash resulting from the explosion of a primordial black hole would be much shorter than that, less than a millisecond, with energy in the extreme upper reaches of the gamma-ray range. A group at the Goddard Space Flight Center in Maryland continues to look for such distinctive flashes, using the EGRET (Energetic Gamma Ray Experiment Telescope), the highest energy instrument on board the Compton.

Dark matter and the clue of lensing effects

Back in the 1930s, the Swiss astronomer Fritz Zwicky was intrigued to discover that galaxies in the great cluster in the constellation Coma Berenices are moving too rapidly, relative to one another, to be held together by their mutual gravitational attraction.

Given everything we know about the way gravity operates, and given what we observe of this cluster of galaxies, the whole arrangement ought to be flying apart, but it isn't. Attempting to explain this anomaly, Zwicky came up with two possibilities. The first was that what appears to be a cluster might instead be a short-term random grouping of galaxies. Zwicky's second suggestion was more unsettling. There might be a great deal more to these galaxies than meets the eye or the telescope. In order to provide the amount of gravitational attraction needed to bind the galaxy cluster together, the cluster would have to contain much more mass than we observe. There was, of course, a third, even less acceptable possibility: we might have made a colossal error in figuring how gravity operates or in assuming that it operates the same everywhere in the universe.

Zwicky's discovery gave birth to the startling notion that we may be able to observe only a tiny fraction of all the matter in the universe. Plenty of support, both theoretical and observational, for the existence of 'dark matter' has emerged in the years since he first speculated about it. We have been forced to conclude that for everything to work as it does, there has indeed got to be much more matter in the universe than we detect with our present technology. It would seem that 90 to 99% of the matter in the universe is not radiating at any wavelength.

For example, something much closer to home than the great cluster in the constellation Coma Berenices: we know that the mass and distribution of observable matter in our own galaxy cannot account for the way the galaxy rotates. What would it take to cause the Milky Way galaxy to rotate as it does? If the matter of the galaxy were mostly outside the visible disk of the galaxy, if it were to extend well beyond what we see as the edge of the disk and if much of it were not level with the disk at all but 'above' and 'below' it, then the rotation would make sense. It would seem that the galaxy must be surrounded by a halo of dark matter – a halo that is much larger than the observable mass of the galaxy. We are not talking about a little fringe around the edges here. The Milky Way galaxy as we know it is about 100,000 light years in diameter. According to studies in 1993 of the orbit of the Large Magellanic Cloud, a satellite galaxy of the Milky Way, the total diameter of

Figure 9.1. (DENNIS THE MENACE, used by permission of Hank Ketcham and © by North America Syndicate.)

the galaxy with the dark-matter halo might be 500,000 light years in diameter or more.

The mystery of dark matter is particularly significant because it impacts upon our understanding of the origin of the universe, its age, and its ultimate fate. Scientists in the early 1990s received a lot of ribbing for coming up with ages for stars that are greater than the supposed age of the universe. But, to be fair, calculating the age of the universe and its life expectancy is an extremely complicated undertaking. Right now there is no consensus as to the rate at which the universe is expanding (the Hubble constant). There's as much support for low values as high ones, which means that estimates for the age of the universe range from 8 billion to 20 billion years, quite a discrepancy. But even were we to pin down the Hubble constant to everyone's satisfaction, there is more to figuring the age of the universe. The age estimate depends also on which model of theory one prefers, whether the expansion has always been the same, and (of most relevance to us here) the density of matter and energy in the universe.

It stands to reason that even with the rate of expansion in dispute, we might be able to add up the voting power (as Wheeler would think of it) of all the matter in the universe and be able to calculate at least *roughly* whether that will produce enough gravitational attraction to stop the expansion and close the universe. When we do this calculation, we find that the amount of matter in the universe that we can observe directly with our present technology is not sufficient to stop the expansion. Case closed, universe open? Not necessarily . . . in large part because of the unknown quantity of dark matter.

Today, sixty years after Zwicky first speculated about it, we still don't know how much dark matter there is and what it is. Candidates range from exotic subatomic particles weighing 100,000 times less than an electron to individual black holes a billion times more massive than the sun, with primordial black holes, planets, and dead or failed stars ('brown dwarfs') making up a broad middle ground of possibilities. We do know that the dark matter cannot all be in the range between the size of the planet Jupiter and the size of a brown dwarf. Present estimates say that only about 20% of it is in that range.

As with black holes, we study dark matter by noting how it affects other things. We observe its gravitational effect on other matter and radiation. Dark matter might reveal its presence not only by the fact that it's needed to explain how galaxies and galaxy clusters hold together, but also by the way it bends the paths of light. As we noted in earlier chapters, the paths of light through

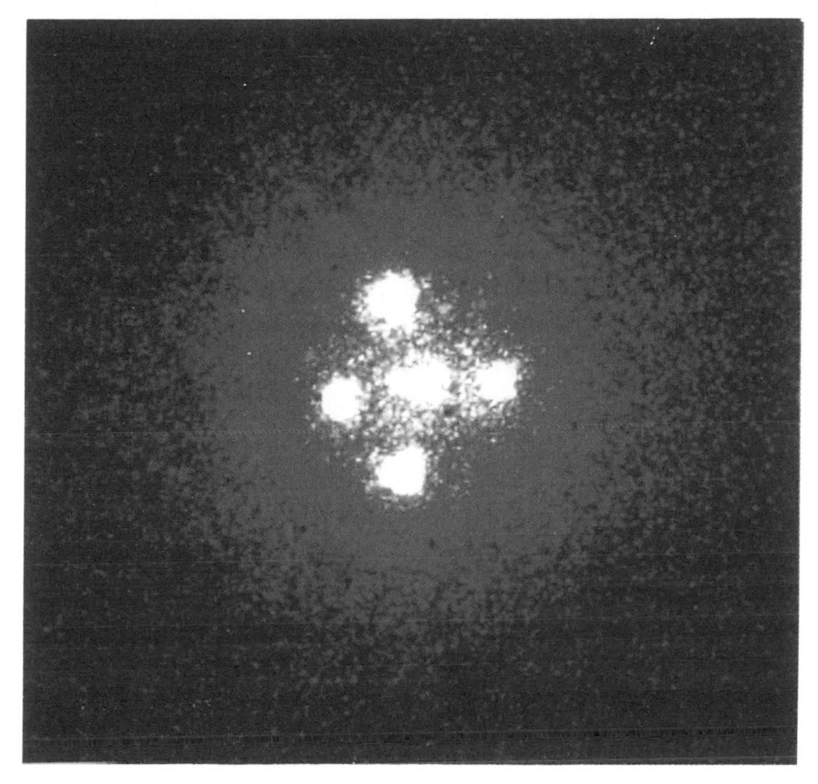

Figure 9.2. The 'Einstein Cross'. Paths of light from a distant quasar are bent as they pass a galaxy much closer to us. As a result we see four images of the quasar (there is perhaps a fifth hidden by the galaxy) distributed around the image of the lensing galaxy. (NASA/ESA.)

spacetime are bent by the presence of massive objects, regardless of whether or not these massive objects – these 'benders' – are themselves detectable at any wavelength. There are many instances of such 'gravitational lensing'. Certainly not all of the examples relevant to a meaningful measurement of the density and expansion rate of the universe involve black holes or dark matter. There are multiple images of quasars created by massive intervening galaxies (Figure 9.2). There are luminous arcs and rings that we now interpret as distorted images of galaxies located behind galaxy clusters. However, when it is possible to tell that the distortion is too great to be caused by the observable matter in the 'bender', or when we cannot directly detect any bender at all, we know we are not seeing everything that's out there between us and the distorted background. We suspect the presence of dark matter.

Let us review the way gravitational lensing works. There are three components to this operation: the background source of radiation, the observer (us), and, between the two, the bender – whatever massive body or collection of mass it is that is acting as a gravitational lens. In Chapter 6 we saw how a gravitational lens (in that case, a black hole) might distort our image of the background (Figures 6.1 and 6.2). There is an additional effect: the lens can make the source appear brighter, because light that normally wouldn't strike the earth at all is redirected toward us.

While galaxies and clusters of galaxies can act as lenses, so can individual objects within galaxies and clusters – a phenomenon known as microlensing. It is reasonable to assume that stars, black holes, and chunks of dark matter within our own galaxy might cause rather rapid changes in the appearance and brightness of background sources as we observe them from earth. Unless a very massive black hole were the bender, these changes would most likely be noticeable to us only as changes in brightness. How long the change lasts would depend on the relative distances and movements of the background source, the lens (bender), and ourselves, and on the lens's mass – which gives us a potential method for finding and calculating the mass of dark matter that may exist as star or planet size objects in our galaxy.

It's fortunate that we have not just one but two rich background areas to provide a great many background sources against which a gravitational lens in our galaxy might reveal itself: the galactic centre and the Large Magellanic Cloud (LMC). Light reaching earth from the LMC does not have to pass through the dense stellar population of our galaxy's disk or central bulge. It crosses the relatively empty region surrounding the visible galaxy, the halo region believed to contain a good share of the galaxy's mass in the form of dark matter. If there are planets, black holes, or dead or failed stars in that halo region, one or another of them should occasionally cross the line of sight between us and some star in the

LMC, and, if we are watching with a telescope, we should notice the change in the brightness of the background star.

The idea that we ought to be looking for such changes in brightness originally came from Bohdan Paczynski of Princeton University and was calculated in more detail by Kim Griest of the University of California, San Diego. They calculated how frequently lensing events should occur for different lens masses in the directions of the galactic centre and the Large Magellanic Cloud. Events last longer toward the LMC because of its greater distance from us, and, in general, lensing events by more massive objects are less common but last longer – which means that black holes are more difficult to detect than less massive candidates for dark matter. It might at first seem that longer duration would make detection easier, but that isn't the case. You have to be watching longer with the telescope to catch the change in brightness. For instance, nightly observations could detect lenses with 0.001 to 0.005 solar masses, while it would require an experiment lasting a few years to detect a lens with many hundred solar masses.

Efforts to find instances of microlensing began in 1990 and there are now several groups involved in the search. So far there have been results suggesting lenses with masses probably too small to be black holes resulting from the collapse of stars and far too large to be Hawking's primordial black holes. Deepening the mystery of dark matter, observers have found many fewer events than expected in the galactic halo and more than expected toward the galactic centre. However, what is clear from these studies is that microlensing does indeed occur, providing a new way to explore the nature of dark matter and perhaps to discover the presence of black holes.

Detecting gravity waves

You'll recall from Chapter 6 that after a star has collapsed to form a black hole, there is very little we can find out from that black hole concerning the star it used to be. 'Black holes have no hair.'

Figure 9.3. A diagram depicting the outward moving ripples of spacetime curvature, called gravity waves, around a binary system made of two black holes. (Courtesy of the LIGO project.)

It would seem we can't even figure out by studying a black hole whether it resulted from the collapse of a star or from the collision of two smaller black holes. However, it may turn out that the provenance of a black hole is not so entirely unknowable after all. Some record of its history may be encoded in ripples of spacetime curvature – gravity waves.

In Chapter 4, we were warned that gravity waves might be a danger to a ship and its passengers stationed near a collapsing star. Now we are about to see that they could also be our best hope for studying real black holes in the universe from where we are on the earth.

To recap: gravity waves are travelling ripples of spacetime curvature, analogous to travelling ripples in the surface of a pond. Einstein's general theory of relativity predicts that gravity waves are produced whenever two black holes or two stars orbit one another (Figure 9.3), when they collide, when a star whose distribution of mass is not spherically symmetric collapses to form a black hole, when objects fall into a black hole, and at the Big Bang origin of the universe.

Objects and events we could study with gravity wave detectors might be pretty much invisible to instruments that study electromagnetic radiation. The 'black hole signature' astronomers and physicists have been hoping to find may very well be written in gravity waves. As the curvature ripples travel outward in spacetime, they are not distorted by intervening matter, as light, X-rays, and radio waves are. Furthermore, their origin is far closer to a hole's horizon than the origin of the other radiation astronomers have been studying to gain understanding of black holes. Most significant, gravity waves are made of the very stuff out of which black holes are made – spacetime curvature.

The trick is to detect the ripples of spacetime curvature and decode the message in the ripples as these ripples pass us. Although their strength where they originate is prodigious, likely to cause a stretch and squeeze that could easily kill a human being (see Chapter 4), they are very weak indeed by the time they reach the earth, and almost impossible to detect.

The effort to build an instrument that can detect these weak waves began in the 1950s and continues today with increasing promise of success. However, the first proof that gravity waves *do exist* did not have to wait for the invention of these instruments. It came from observations of Joseph Taylor and Russell Hulse of Princeton in 1974, for which they won the 1993 Nobel prize. To understand Taylor and Hulse's discovery, you must know that when gravity waves start out, there is a 'kick' analogous to the way a rifle kicks when you fire it. In a case where two black holes or two stars orbit one another, the gravity waves kick back in this manner on the holes or the stars, driving them closer together and increasing their orbital velocities. The resulting inward spiral of the two objects releases gravitational energy, half of it going into gravity waves and the other half into increasing the objects' orbital velocities still more. What is a slow spiral at first, speeds up as the objects draw ever closer to one another. The faster they move, the more strongly they radiate ripples of spacetime curvature, and the more quickly they lose energy and spiral toward one another. Einstein's laws predict the rate at which all this should occur. Taylor and Hulse, with a radio telescope, discovered two neutron stars, one of them a pulsar, orbiting each other in precisely the way

Einstein's laws indicate they should. There is no other explanation for the way these stars are spiralling toward each other except gravitational wave 'kicks'.

10

Passages into the labyrinth

Stephen Hawking of Cambridge University, the great British theoretical physicist who discovered Hawking radiation, lecturing on wormholes and baby universes at Northeastern University in Boston. (AP/World Wide Photos.)

listen, there's a hell of a good universe next door:
let's go!
e.e. cummings

Whenever conversations turn to black holes, someone is likely to ask whether black holes might provide a way to travel backward and forward in time and to parts of the universe we otherwise could never reach. Leaving aside questions whether such travel would be a good thing or what risks there might be of our meddling in the past or the future, we have to admit that this would be an exciting and perhaps even very practical use for black holes. If it works, we can wonder why tourists from advanced past, present and future civilizations haven't already shown up on our shores with their phrase books and cameras, asking whether it's safe to drink the water. Stephen Hawking comments about the possibility of time travel that the noticeable lack of such visitors indicates either that we live in a period of history no one cares to visit, or else that time travel will never be possible by any method. Of course there is also the unsettling thought that the tourists might be consummate masters of disguise.

Most physics theorists today, with a few exceptions, have regretfully concluded that it is not possible to use a black hole's core as a passage to another place or time in the universe. We have already seen that anything falling into a black hole is almost certainly torn apart and crushed out of existence before it reaches the core of the hole. Even if we did somehow make it that far, we would be mercilessly pummelled by a hail of electromagnetic vacuum fluctuations (see Chapter 5) and tiny amounts of radiation that continuously bombard a black hole. A black hole's gravity accelerates these to ultra-high energy as they fall. Any vehicle for hyperspace travel, any astronaut no matter how good his or her protective clothing, would be destroyed by this explosive rain before the trip to another part of spacetime even got under way.

Are 'wormholes' a better bet for space and time travel? We see them used frequently these days for short cuts through space and time in science fiction stories and TV space dramas. 'Deep Space Nine' featured one, and the 'Star Trek Voyager' crew hoped to find one that would take them back home to a familiar part of the galaxy. Kip Thorne worked with Carl Sagan to make travel through a wormhole scientifically plausible in Sagan's novel *Contact.* In Chapter 11 we'll weigh the possibility of such journeys and look at some suggestions from Thorne and others. First, in this chapter, a brief overview of wormhole theory and how wormholes are related to black holes.

A wormhole would be like a tunnel from one part of spacetime to another – a tunnel with two mouths, each of which is a sphere resembling a black hole. (See the screen behind Hawking in the frontispiece to this chapter.) However, there is a difference. While the event horizon of a black hole is a one-way street, the mouth of a wormhole need not be. It is hypothetically possible to travel back and forth through it. Light would be able to travel back and forth, so that if the other mouth of the wormhole were near a star or a lightbulb, the light from that star or lightbulb would emerge from the spherical mouth at our end. Though the mouths of the wormhole could be seen from our universe, the tunnel leading from one mouth to the other could not. It would be a passage through hyperspace, not through space as we know it. What makes travel through a wormhole for either astronauts or light very 'iffy' is the fact that according to most theories wormholes would have a life span far too short for anything to make it the distance from end to end before the wormhole pinched off and disappeared. Even worse news: anything entering a wormhole – even bits of radiation – would trigger it to pinch off more quickly.

Wormholes are not firmly established in theory in the way black holes are. There is no known process (such as the collapse of a star in the case of a black hole) by which they could occur naturally in the universe. Furthermore, no one has found any observational evidence of their existence – not even indirect, circumstantial evidence.

Why, then, are we devoting a separate chapter to wormholes rather than discussing them only among 'Wild Ideas' in Chapter 11? It's true that wormholes are in the borderlands between science and science fiction, but physics theorists take them seriously. Some believe wormhole theory may offer insight into the origin of the universe and perhaps help solve some of the most puzzling enigmas in physics. Whether or not things the size of human beings or their spaceships can travel through them, theory suggests that wormholes might be connections to other times and places in our universe and to other universes.

The idea of wormholes is not new. They were discovered as a solution to Einstein's field equation in 1916 shortly after Einstein produced the equation. In the 1950s, a research group led by John Wheeler studied them, and Martin Kruskal, an associate of Wheeler's at Princeton, found a solution to Einstein's equation in which the evolution of a wormhole begins with two singularities in different parts of the universe. These singularities are not like the singularity in a black hole. They are like the Big Bang singularity in which the universe is presumed to have originated – singularities where time flows out and things are created. The two singularities reach out through hyperspace, meet each other and annihilate each other, creating a wormhole. The wormhole gets wider, then slims down again, then pinches off, leaving two singularities. These final singularities are like the Big Crunch singularity in which the universe may end – singularities where time flows in and things are destroyed, including anything trying to travel through the wormhole.

In the 1980s wormholes became a preoccupation of Stephen Hawking and of Sidney Coleman of Harvard, both of whom are particularly interested in the possibility that wormholes are part of the process in which new universes come into being. The wormholes that interest Hawking and Coleman most are extremely small, only about 10^{-33} centimetres across – quantum wormholes. Written out as a fraction, that is 1 as the numerator and 1 followed by thirty-three zeros for the denominator. From our point of view, these tiny holes flicker into existence and then vanish after an interval too short to imagine. They exist not in 'real' time, as we normally experience it with a well-defined past and future, but in 'imaginary' time. Imaginary time is a situation resulting from quantum effects in which the fourth dimension – time – ceases to exist as 'time' and becomes a fourth space dimension. Chronological time is not there at all. 'Past', 'future', 'before', 'after' – all words that are meaningful in the context of chronological time – have no meaning.

Hawking asks us to picture a balloon, an enormous one, inflating rapidly. This represents the cosmic balloon, our universe. Picture dots on the balloon's surface, representing stars and galaxies. These massive objects are curving spacetime as Einstein predicted, and so we can picture them causing tiny dimples and puckers in the surface of the balloon. In some cases the dimples are so deep and severe that the surface has almost pinched them off. These are black holes. However, in spite of these dimples and puckers, the balloon's surface is relatively smooth, even when we examine it through a microscope. We must look at it much more closely, through a much more powerful microscope (one that doesn't exist in our present technology), to see that it is not smooth after all. The surface at this magnification seems to be vibrating furiously, creating a frothy foam.

This description shouldn't seem entirely unfamiliar. What Hawking is asking us to visualize is the quantum level of the universe, where the uncertainty principle causes the universe to be a blurry affair. Recall from Chapter 6 that the uncertainty principle means that a field, such as an electromagnetic field or a gravitational field, cannot have a definite value and a definite rate of change over time simultaneously. Zero is a very definite measurement, so a field's value and rate of change cannot both measure zero simultaneously. In order to have empty space, all fields would have to measure exactly zero in both value and rate of change. No zero, no empty space. Instead of empty space we have a continuous fluctuation in the values of all fields, a wobbling a bit toward the positive and negative sides of zero so as not to *be* zero. When we say there are fluctuations like this in a gravitational field, that is the same as saying there are fluctuations in the curvature of spacetime. These are not big, smooth curves like swells on the ocean. They are all sorts of continuously changing crinkles, ripples, and swirls. At this level, as we scrutinize a very tiny region, the 'surface' of our balloon won't look like a surface at all. It will look more like the 'surface' of a bubble bath (see Figure 10.1). What we are seeing is the quantum foam we found at the core of a black hole in Chapter 5.

According to Hawking and some of his colleagues, in quantum theory what is not forbidden can and will occur, and so it is not surprising to hear Hawking say that under high enough magnification the quantum fluctuation becomes such that there's a probability we'll find it doing 'anything'. Specifically, he thinks there is a probability that the cosmic balloon will develop a tiny bulge in it. I saw something like this happen once as I inflated a party balloon for my son. The bulge didn't cause the balloon to burst, but instead became a miniature balloon attached to the surface of the larger balloon by a narrow neck. If you or I were able to observe this happening with the cosmic balloon, we would be witnessing the birth of a baby universe. The tiny neck would be the wormhole. At our end it would look like a little black hole.

Are there any data to support this theoretical speculation? There have been several proposals for experiments. Hawking himself is pessimistic about the possibility of any of them showing whether or not there really are such things as wormholes. Given their size and the fact that wormholes are supposed to exist only in imaginary time, any data that may emerge will not be direct observational data.

Though the wormhole connection is in imaginary time, the baby universe attached to this umbilical cord may not continue to exist only in imaginary time or stay small. It may expand to become something like our own universe, extending billions of light years. Perhaps not just something like ours. Perhaps exactly like ours, with galaxies, stars, planets, life, black holes. The suggestion Hawking makes is that our universe did originate as a baby universe bulging off the side of another universe. According to the theory, there may be many universes, a never-ending labyrinth of them, connected by wormholes in more than one place (Figure 10.2).

From the point of view of an electron, the universe must look like an enormous, furiously boiling pot of thick, foamy porridge. Theory tells us that an electron moving in such an environment is likely to encounter a wormhole, fall in, and go shooting off into another universe. Will this be a violation of the law of physics that

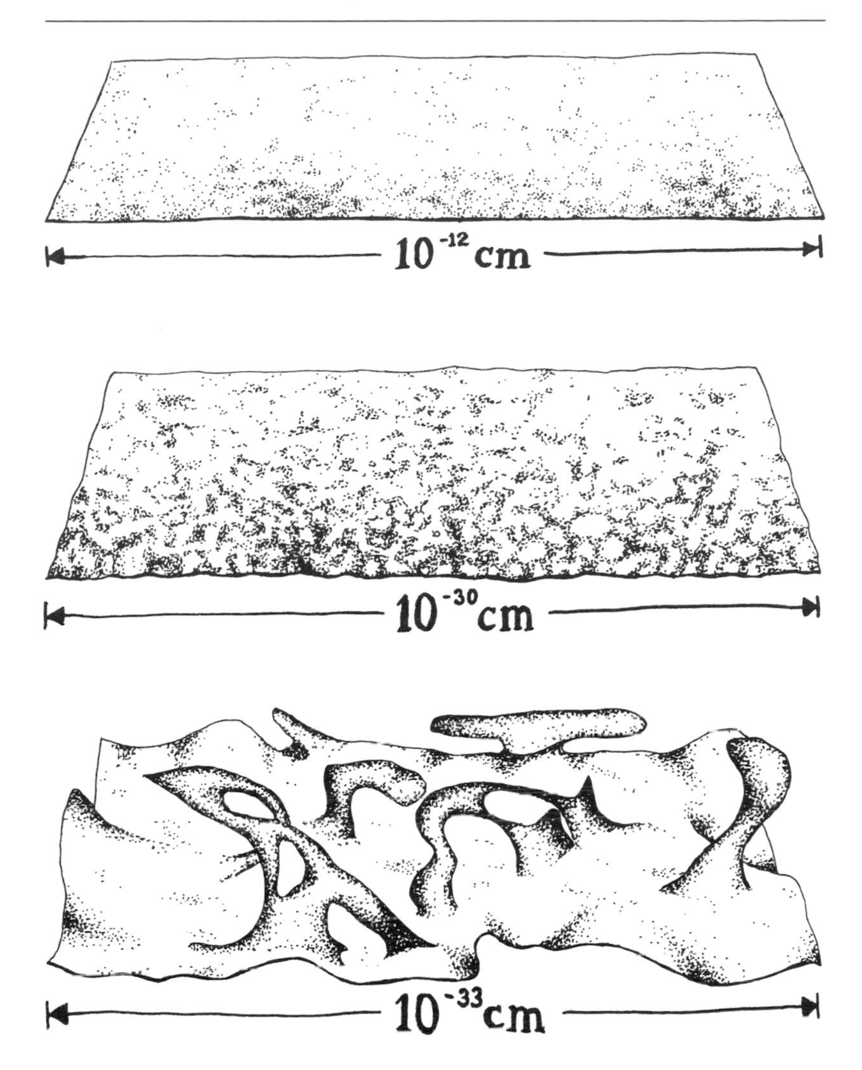

Figure 10.1. The quantum vacuum, as imagined by John Wheeler in 1957, becomes more and more chaotic as we inspect smaller regions of space. At the scale of the atomic nucleus (top) space still looks very smooth. Looking much more closely than that (middle), we see a roughness begin to appear. At a scale 1000 times smaller (bottom), the curvature undergoes violent fluctuations. (Courtesy of John A. Wheeler.)

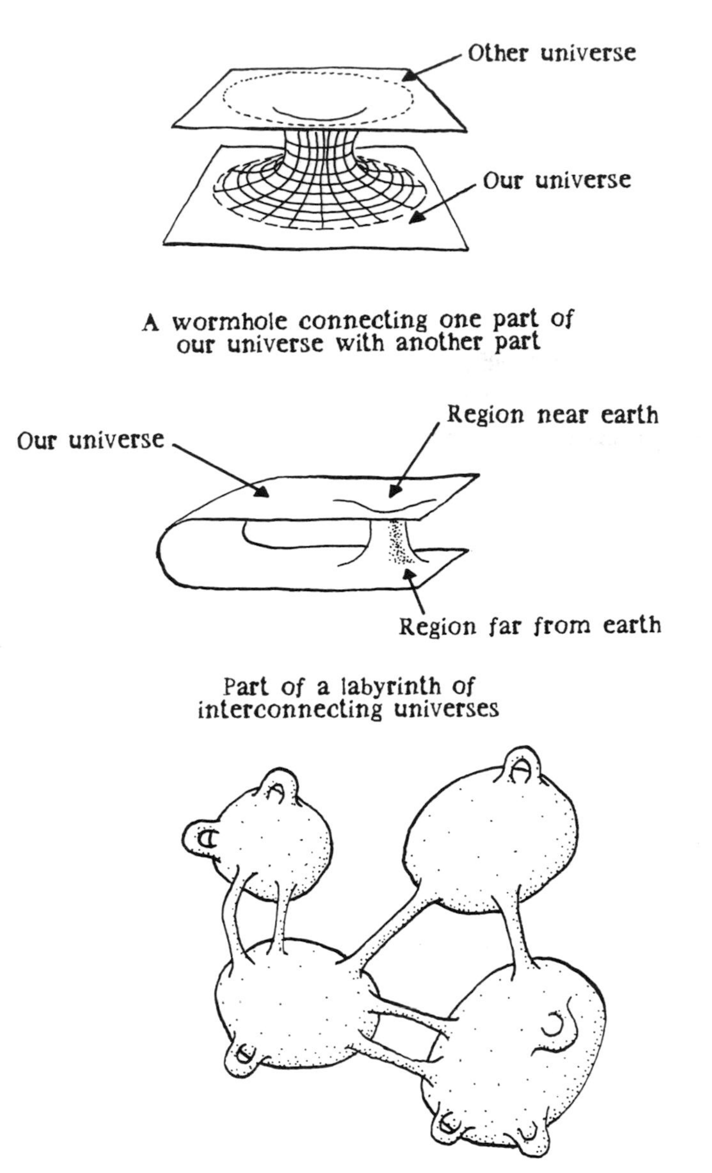

Figure 10.2. Wormholes and baby universes. (Drawn by Andrew Dunn.)

says matter is not allowed to disappear from our universe? No, because an identical electron comes back the other way and pops into our universe. The net loss is zero. To us this event will not look like a substitution but like one electron moving in a straight line. However, the presence of wormholes will have an effect on the mass of all electrons. Therefore, if we want to understand why an electron has the mass it does, it is important to know whether or not there really are such things as wormholes. Electrons are not the only elementary particles that may travel in this manner. Hawking suggests that all particle masses and all particle interactions all over the universe may be explained as this going into and out of wormholes.

Wormhole and baby universe theory makes a stab at resolving one disturbing inconsistency in modern physics. Though relativity theory and quantum theory are both highly reliable theories, put together they do not predict a universe like ours. We have seen that the uncertainty principle of quantum mechanics doesn't allow empty space to be empty. Instead, there is a continuous fluctuation in the value of all fields, a wobbling a bit toward the positive and negative sides of zero so as not to *be* zero. The upshot is that empty space instead of being empty must teem with energy. The energy density of the universe ought to be enormous. Meanwhile, the theory of relativity tells us that the presence of matter or energy causes spacetime to curve.

Putting two and two together. The vacuum seethes with energy; the presence of matter/energy causes spacetime to curve; the more matter/energy, the greater the curving. Mathematical calculations, not to mention common sense, tell us that all that energy in the vacuum ought to have curled the universe up into the size of a small ball. Or, more difficult to understand, if the 'cosmological constant' (the term physicists use to refer to the energy density of the vacuum) is positive rather than negative, the universe would have expanded at such a rate that galaxies couldn't have formed. In either case, we wouldn't be here.

Fortunately, neither has happened. In fact, the value of the cosmological constant is observed to be near zero. Do all the positives and negatives there in the vacuum really cancel one another out that handily? Coleman tells us how unlikely this is: 'Zero is a suspicious number. Imagine that over a ten-year period you spend millions of dollars without looking at your salary, and when you finally compare what you spent and what you earned, they balance out to the penny.' For the cosmological constant to balance out to as near zero as it seems to is even less likely. That value would have had to be set in the very early universe with a precision that defies understanding.

How might wormholes help us solve this enigma? Coleman explains it like this. Imagine the birth of a universe – a baby branching off from an existing universe. According to the theory there may be plenty of universes around, some more enormous than ours is today, others unimaginably smaller than an atom, and all sizes in between. The newborn universe must copy its cosmological constant value from one of these other universes through a wormhole attachment – 'inherit it', you might say. It's of little importance to a human infant whether it inherits a talent for music; it becomes important only when the infant grows larger. It is of little importance to a baby universe whether it 'inherits' a cosmological constant value near zero or one which would curl it up into a small ball. Its cosmological constant value will not even be measurable until it's quite a bit more grown up. However, with all those myriad assorted sizes of universes around, the infant has a far better statistical chance of inheriting its cosmological constant value through wormhole attachments with large, cooler universes of the sort possible only when all those positives and negatives in the vacuum do cancel out to near zero. Coleman calculated the probability of a universe (in wormhole theory) being a universe where the cosmological constant is near zero: our sort of universe. He found that any *other* sort of universe would be highly unlikely.

Can wormhole theory help explain some of the other constants of nature – such as the masses of particles and the strength of the gravitational force – that we can observe but cannot predict with any of our theories? Many of these values seem to have been inexplicably fine-tuned at the origin of the universe in a way to allow for the universe to evolve into what we observe today and for beings like ourselves to emerge. No, this theory doesn't work such magic with other constants. Coleman and Hawking tell us that it might if we were able to map the entire labyrinth of wormholes and universes, but of course we can't do that nor will we ever be able to. We can conclude, however, that if these theorists are right, the reason why these constants of nature are knowable only through observation and can't be predicted by any theory is that they arise from a situation in which chance plays a part, leaving us at best calculating probabilities, not with exact predictions.

Could you or I ever travel through one of these quantum wormholes into other universes – or explore the labyrinth of universes? Not unless we are the size of a fundamental particle and can function in imaginary time. As for travelling through larger wormholes, we will leave discussion of that for the next chapter.

11

Black hole legends and far out ideas

Suddenly, through forces not yet fully understood, Darren Belsky's apartment became the center of a new black hole.

I'm entering the black hole now! . . . Good grief! . . . It's full of unmatched socks!

from a cartoon by 'Chase'

We now come to the most fanciful chapter in this book, though some might think it the most useful. It's fanciful in that we shall allow ourselves to stray occasionally beyond the bounds of accepted theory, useful in that readers will find herein a wealth of interesting conversational possibilities. Let's face it, if word gets around that you have become something of a mini-expert on black holes, will your friends ask you questions that allow you to show off that you know at what circumference the event horizon forms? No. They will ask 'Is there a black hole in the Bermuda Triangle?' or 'Is the Loch Ness Monster going in and out of a wormhole?' This chapter will help you with social situations like that.

Everything you have read so far in this book is well-supported by current theory. At the boundaries of theory, and beyond, there are some wild rumours going around . . .

No black hole at Tunguska

What if a black hole were to collide with the earth? For a while there was serious speculation that such an event had occurred. Books about black holes published in the 1980s had illustrations showing the results of the disaster.

Here's the story.

On the morning of 30 June 1908, a fireball passed across the sky of China and Russia. When it was about 8 kilometres above the ground, there was an explosion so powerful that it disturbed the magnetic field of the earth. Witnesses a thousand kilometres south-east saw a cylinder of fire rising far into the sky, and buildings trembled. The sky in northern Europe was so strangely brilliant for many nights that Londoners telephoned the police to inquire whether the northern part of the city was on fire.

The centre of the blast was in an extremely remote area on the Central Siberian Plateau, near the Stony Tunguska River, an area frequented only by reindeer herders. It wasn't until nineteen years after the event that an expedition slogged through the icy swamps and dense forest to reach the site.

They found incredible devastation. Trees within a 30 to 40 km radius ripped up and laid out like charred pickup sticks with their dead roots pointing to the centre of the blast area; trees further away blackened and stripped bare; the ground scorched and riddled with unexplainable holes and ridges – but no crater.

Speculation was rampant and has been ever since.

A meteorite? That first expedition thought it must have been, though they found no crater and no metallic remnants of the sort usually left by a meteorite, and meteorites do not usually explode just before reaching the ground.

An asteroid? It seemed that an asteroid big enough to have produced so powerful a blast would have partly survived and hit the ground, leaving some trace or crater.

A comet? Surely that would have been spotted earlier approaching the earth. Comets don't exactly sneak up and take us by surprise.

Antimatter? A much more exotic suggestion! Matter and antimatter, when they meet, annihilate one another in an explosive fashion. But how would any hunk of antimatter have avoided meeting matter much higher in the atmosphere than this explosion occurred?

A crippled alien spacecraft? Perhaps the pilot knew its nuclear-powered engines were about to explode and deliberately plotted a course so that the crash landing would occur in an area of our planet where it would cause least harm.

A primordial black hole? Speculation had it that as the tiny black hole approached the ground it would have created a shock wave in the air great enough to produce temperatures from ten thousand to a hundred thousand degrees Celsius. The radiated light, absorbed and reradiated by air along the path, would have been the blinding pillar of fire. The heat and radiation would have caused the burning and charring. The blast would have blown down the trees and shaken the earth, even though the black hole itself did not explode. Being smaller than an atom, the black hole would have passed right through the earth. On the opposite side, it would have emerged in open ocean, causing an atmospheric disturbance and an enormous geyser of water as it headed off into space once more.

Well . . . why not? It sounds intriguing and not entirely ridiculous. Unfortunately, the arguments against blaming a black hole for the Tunguska disaster are now fairly conclusive. There was no recorded geyser or atmospheric disturbance. In the 1980s more careful combing of the site yielded the sort of debris expected from a meteor. Recently, scientists from NASA and the University of Wisconsin came up with what they believe is the most satisfactory explanation yet: a 30-metre stony asteroid. A stony asteroid is the most common type of asteroid, but this one had to be just the right size. A larger stony asteroid would have hit the earth. A smaller one would have exploded much higher than the Tunguska explosion took place. A 30-metre stony asteroid could have exploded into tiny fragments at a height of about 8 kilometres.

One really would *like* a more romantic explanation, but there we are.

The matter eater

Let us suppose a primordial black hole the size of an atomic particle did collide with the earth. Let us further suppose that instead of passing through the earth and heading off once more into space it oscillated back and forth through the earth again and again, eventually settling down at the centre. There it sits, a tiny black hole the size of an atomic particle. What now?

You might think that such a black hole would become a 'matter eater'. First it would suck in particles near it. That would increase its mass and extend its event horizon, putting it within range of still more particles – its next victims. It would eat and eat and grow and grow until the entire earth was swallowed up.

That has not happened . . . so there must be no primordial black hole in the centre of the earth . . . or perhaps the black hole is a slow eater. Physicists are in disagreement. When asked how long it would take a primordial black hole to consume the earth, some answer 'Not long!'. Others, recalling all that empty space within atoms, say, 'Probably far longer than the future of the solar system, perhaps even longer than the future of the universe.'

Meeting ourselves coming and going

Marek Artur Abramowicz of the University of Göteborg in Sweden, with Jean-Pierre Lasota, now at the Paris Observatory, describe the following in an article in the March 1993 issue of *Scientific American*. Though the idea may sound like science fiction, there is nothing dubious about it. It follows logically from what we know about black holes, and about gravity and light. Building the apparatus to experience it . . . well, that is another matter.

We construct a circular tube around a black hole, above the event horizon at the exact elevation at which the curvature of spacetime bends light rays in a perfect circle centred on the black hole (Figure 11.1).

We know the tube forms a circle because we built it that way, we have seen it from a distance, and we've measured the curvature of the walls along the tube using straight rulers. However, standing in the tube, our eyes tell us differently. This tube is *straight*.

We set up a lamp in the tube and walk away from the lamp. Knowing the tube is circular (in spite of what our eyes tell us), we expect after having walked a little way to look back and not see the lamp. It will be out of sight around the bend. Instead, no matter how far we walk around the tube, we continue to look over our shoulders and see the lamp straight behind us standing in the middle of the tube, though more and more distant. What is more, we also see the lamp straight ahead of us, less and less distant as we walk toward it.

It should not be any problem for you, having read the previous chapters, to understand what's causing this paradoxical situation.

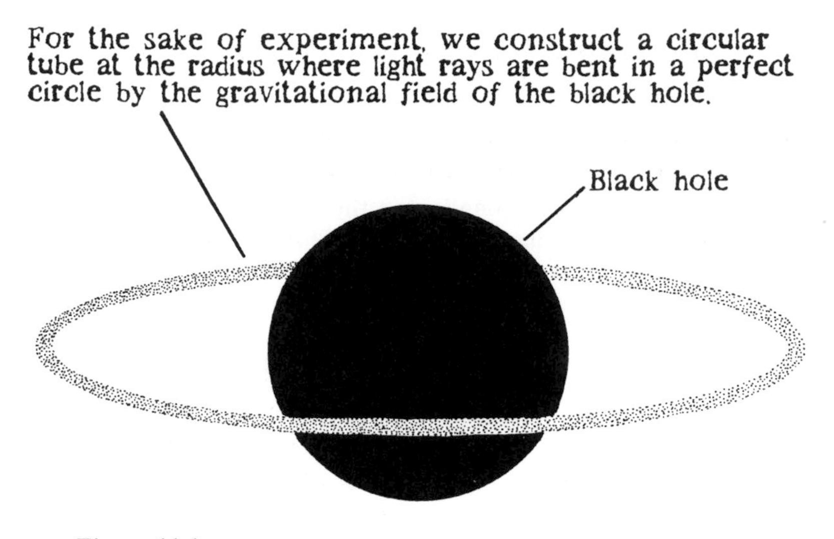

Figure 11.1.

The path of light is bent by the curvature of spacetime around the black hole, and we instinctively judge light to be coming from the direction it enters our eyes. Our eyes and brains have no way of registering any change in the direction of its path prior to that. And so, the image of the lamp travels round and round the tube to meet our eyes no matter where we are. We see, actually, many images of it superimposed one on the other. Wherever we are in the tube the light from the lamp reaches us along the same circular path.

Except for illumination, we don't really need a light to observe this phenomenon. Looking forward down the tube, which looks perfectly straight to us, we see our own backs.

Even more intriguing, if we were to construct our circular tube nearer the black hole, the path of light would be so bent that the tube would appear to curve *away* from the black hole, opposite from the way we know it is actually curving.

Abramowicz and Lasota carry the discussion on from there to show that very close to a black hole we would feel a centrifugal force pushing inward toward the hole rather than outward. For that slightly more complicated argument, I refer you to their article.

Hawking radiation in the clothes washer

Respectable members of the physics community have been heard to blame black holes in their clothes washers for the disappearance of socks in the washing. Hawking's work on particle pair separation at the event horizon sheds new light on this problem. We now understand why so often only one of a pair vanishes.

What must surely happen is that sometimes a sock becomes 'real' in the wash and is able to escape to a distance, while its partner falls into the black hole. But even a 'real' unmatched sock has little value except as a toy for the dog. Also useless, except perhaps for its scientific interest, is an unmatched sock that appears in a household where one of its size, design, colour and odour has never been previously observed. In this case, look for a wormhole.

The mother of all garbage recyclers

Charles Misner, Kip Thorne, and John Wheeler make this proposal in their book *Gravitation* (Figure 11.2).

A civilization much more advanced than our own constructs a rigid framework around a rotating black hole, not at the event horizon or the static limit but quite far out beyond both. On this framework they build a city. Each day garbage trucks carry all the city's garbage to a dump where the garbage is transferred into shuttle vehicles. These are dropped through an opening in the rigid framework, and they fall toward the black hole.

As a shuttle reaches the static limit and enters the ergosphere, it is whipped into a circling, inward-spiralling orbit near the event horizon. When it reaches a certain 'ejection point' the shuttle ejects its load of garbage. While the garbage falls into the black hole, increasing the mass/energy of the hole, the shuttle recoils from the ejection and goes flying out away from the hole with more energy than it had going down.

As shuttles hurtle out through the opening in the rigid framework, they strike a flywheel which turns a generator, producing

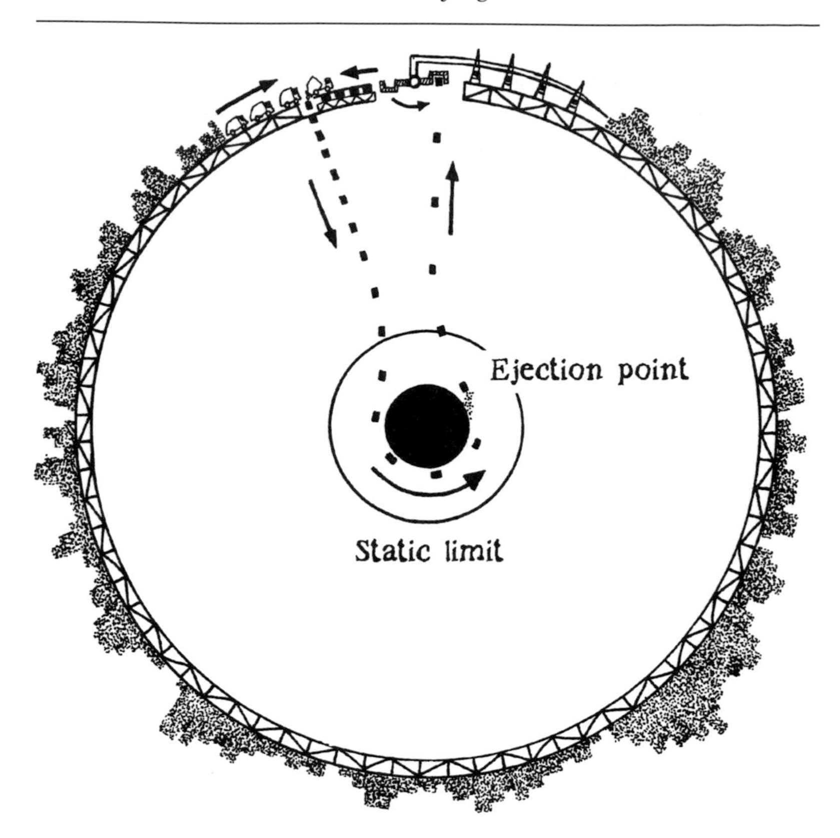

Figure 11.2. A black hole becomes the ultimate garbage recycler. (From GRAVITATION, by Misner, Thorne and Wheeler, © 1973 W.H. Freeman and Co. Used with permission.)

electricity for the city. In this way, the inhabitants of the city use the Penrose Process to convert the entire rest mass of their garbage and some of the mass of the black hole into electric power.

Donkey and carrot

Instead of our moving to a black hole and building a city on a framework around it, suppose we want to stay where we are and bring a black hole to us. One primordial black hole emits enough energy to run several large power stations, so it's well worth considering. We would not, of course, want to have it sitting right on the surface of the earth. In fact, it wouldn't sit there. It would fall right through. It might even end up as a 'matter eater'. A better place for it would be in orbit around the earth. How to get it there? Hawking suggests using a tried and true method: attract it by towing a large mass in front of it, like a carrot in front of a donkey.

Galactic Grand Prix

The ultimate challenge for daredevils of the future! The object of the race: to circumnavigate the galaxy and reach the finish line last and youngest.

The competitors start off in their spaceships. The closer to lightspeed they can travel, and the more black holes they can visit, and the nearer they can skim the event horizons, the more they will benefit from time dilation and the longer they'll extend their lives. They must plot their courses with extreme care in order to avoid lethal tidal effects and not stray across an event horizon. Event horizons, of course, can't be seen. The best pilots will return to earth billions of years after leaving it, still young, athletic, mathematically brilliant, ready to receive the accolades of their remote descendants, who we hope will recall that there's been a race going on.

Black holes and white holes

At a time when quasars were much more of a mystery than they are today, there was speculation that they might be 'white holes', though no one was sure what a white hole was. The time-reversed version of a black hole? The opposite end of a black hole, spewing matter into our universe that had fallen into a black hole in another universe?

Perhaps you will be better able to render judgment on the theoretical respectability of white holes if you know the following story. When Roger Penrose arrived for the 1971 Texas Symposium on Relativistic Astrophysics, he found himself scheduled to give a talk with the unexpected title 'Black Holes and White Holes'. Though Penrose knew very well what black holes are, he wasn't so sure about white holes, so he discreetly inquired among the other participants. As puzzled as he, they nevertheless came up with a variety of suggestions: a time-reversed black hole . . . a 'naked' singularity (one not hidden by an event horizon) . . . a black hole having its rotational energy extracted by means of the Penrose Process . . . and so forth. Not wanting to renege on his title, and seeing that the field was wide open, Penrose 'invented' white holes. He projected a spacetime diagram of a black hole upside down and used a ten-foot-long pointer in order to remain at a safe distance. Then he made a joke that, unfortunately, neither he nor anyone else present can now remember. Maybe they are still chuckling about it in another universe.

Tunnels through spacetime

When physicist and science fiction author Carl Sagan was finishing his novel *Contact*, he sent the manuscript to physics theorist Kip Thorne, asking Thorne's opinion as to whether it was scientifically plausible for the heroine to use a black hole as a short cut to a distant part of the universe. Thorne replied regretfully that it is not possible to travel from the core of a black hole through hyperspace to another part of the universe. On the other hand, said Thorne, after giving the matter some thought, she might be able to travel through a *wormhole* that was not associated with a black hole.

Thorne continued to consider the possibility, at first by himself and later with his students, and he devotes most of a chapter in his book *Black Holes and Time Warps: Einstein's Outrageous Legacy* to the possibility of such tunnels through spacetime.

There are several requirements that a passage like that would have to meet. First, the passage must stay open and not pinch off and destroy whatever is passing through. Second, there must be no strong tidal forces. Third, it must be a two-way street – in other words, not have an event horizon as a black hole does. Fourth, it must be possible to travel through it in a reasonably short period of time. Fifth, it has to have a safe level of radiation. Thorne and his student Michael Morris decided that with a wormhole it would be possible to avoid or overcome these problems. The first and most difficult, 'pinch off', had been addressed earlier by several physicists including Don Page, Dennis Gannan, C.W. Lee and Tom Roman. They had concluded that the way to hold a wormhole open is with 'exotic material'.

We have spoken several times in this book about the vacuum fluctuations that occur at all times and everywhere in 'empty space'. Under normal circumstances on earth, the average energy of vacuum fluctuations is zero, not negative or positive. 'Exotic' material on the other hand would have negative average energy density as seen by someone travelling through it at near the speed of light.

There is one sticky problem for any advanced civilization or science fiction author trying to thread a wormhole with this sort of material: 'exotic' material may not actually exist. Theoretically, however, it does. For instance, theory tells us that a black hole distorts vacuum fluctuations near it so that they are not what they would be elsewhere. They become 'exotic'. It is the fact that these fluctuations are exotic that allows a black hole's horizon to shrink as a black hole evaporates. Vacuum fluctuations in a wormhole might be distorted in a similar way. Could the spacetime curvature of the wormhole make them exotic and enable them to hold the wormhole open for Sagan's heroine? No one knows the answer to that question, but Thorne felt confident enough to advise Sagan to have a character in *Contact* 'discover' that exotic material *can* really exist and that it can be used to hold wormholes open.

Traversing one of these wormholes might get us to a distant part of the universe, but it would not, in itself, send us backward or forward in time. Could we turn a wormhole into a time machine?

Thorne and Morris, along with another of Thorne's students, Ulvi Yurtsever, toyed with the notion that if two mouths of a wormhole started out near one another, and one of the mouths moved off into space at an extremely high speed (for a significant effect, about 99% the speed of light) and then returned to a location near the mouth that had stayed at home, the wormhole would indeed become a time machine. Let's say there is a clock attached to each mouth. When the travelling mouth begins its journey, the clocks read the same. When the travelling mouth returns, and the two ends of the wormhole are near one another again, both clocks show that some time has passed, but the clock at the well-travelled end shows an earlier time than the clock at the stay-at-home end. (Recall our discussion of time discrepancies in Chapter 5.)

There are limits to what we could accomplish with such a time machine. The clock at the well-travelled end could never show a time earlier than when that mouth had begun its travel. That means you and I can't start the process today – launch the mouth today – and journey back to yesterday. However, if we are young, we might launch the mouth on its journey today, live out our lives, and meet the returning end when we are old. The clock attached to that returning mouth will show that only a few hours or days have passed since we launched it, though for us many years have gone by. Hence, we can travel through the wormhole to a time in our youths shortly after we launched the mouth, meeting our younger selves there. Those younger selves, and our older selves, can go on using the wormhole to travel back and forth in time. Or we can hope that we'll discover the mouth of a wormhole launched on its journey ages ago by a forgotten civilization. We could travel back through that to a time shortly after it was launched.

Are there any large wormholes in nature for us, or a civilization much more advanced than ours, to play around with in this way? Just in case there are not, Thorne speculates that we might find a way to reach down into the quantum foam, grab a wormhole, and enlarge it. Or, it might be possible to create a fold in spacetime similar to the one shown in Figure 11.3, then push down a small area of the upper part of the fold (as one might in a piece of elasticized cloth with the point of a pencil), tear a small hole in the end of that push-out, and sew the edges of the tear to edges of a small hole in the lower part of the fold. One wants to ask why a civilization that could travel around spacetime handily enough to

One suggestion comes from Kip Thorne, who speculates that an advanced civilization might create a fold in spacetime (*a*), then push down a small area of the fold (*b*), tear a little hole in the end of the push-out (*c*), and sew the edges of the tear to edges of a small hole in the lower part of the fold (*d*).

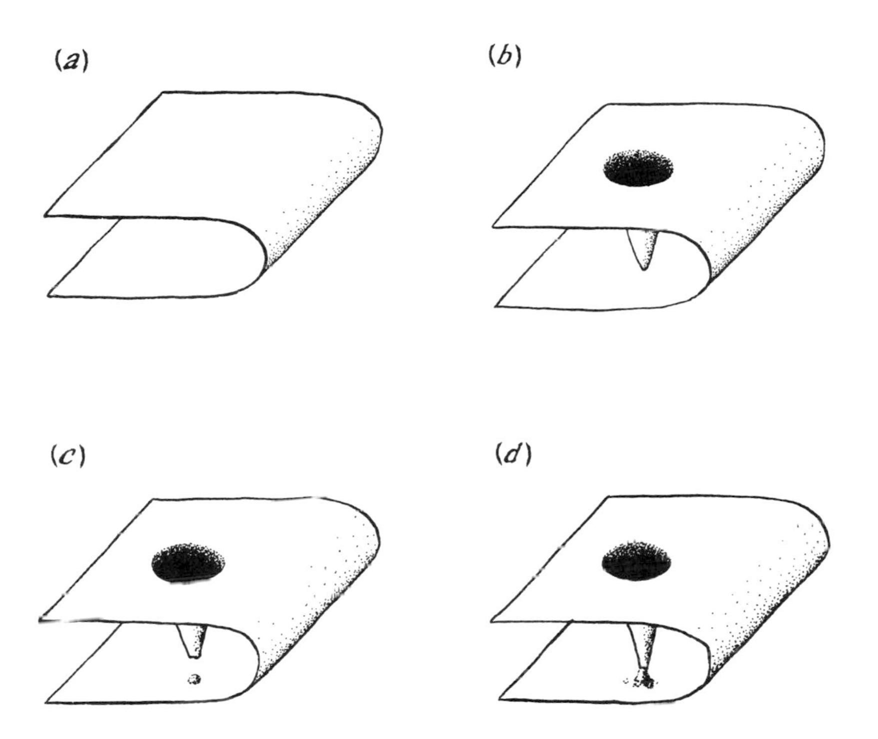

Figure 11.3. Creating a wormhole.

do all this folding, punching, tearing and sewing would need the wormhole.

In 1966, Robert Geroch, then a student of Wheeler at Princeton, showed that it might be feasible to construct a wormhole in another way by a smooth warping and twisting of spacetime without tearing it. The catch is that it would have to be possible to travel backward as well as forward in time during the construction. The apparatus that does the constructing would somehow have to act as a time machine transporting things from late in the construction period back to early in the construction period, but not to earlier than the construction began.

Will any of these ideas ever become reality? No one can answer that at present, nor will they be able to do so until we have a better understanding of how quantum gravity works. Meanwhile, the use of wormholes to travel through space and time is plausible enough so that we needn't sneer when science fiction authors and screenplay writers allow their characters to journey safely through a wormhole, nor should we assume that theoretical physicists have abandoned respectable physics when they study such possibilities. Far from it.

We haven't mentioned the Bermuda Triangle and the Loch Ness monster. Having read this book carefully, you should be capable of fielding those questions on your own!

Epilogue

The most compelling motive for studying black holes has never been anything so mundane as the prospect of developing power stations or garbage disposals, nor has it been anything so fanciful as competing with the White Queen in believing the impossible – though that is perhaps nearer the mark. What has drawn some of the world's best minds to the subject of black holes is the belief that exploring these remarkable objects will bring us closer to Wheeler's 'deep, happy mysteries', to the principles that may, with astounding simplicity, underlie all the diversity and marvel of this universe.

The evolution of our knowledge about gravity and black holes has been a gradual journey to a more profound understanding of nature. Newton's theory of gravity was one of the greatest intellectual insights in human history, yet we found early in the twentieth century that these laws cannot account for what happens when movement approaches the speed of light or when gravity becomes enormously strong. Einstein's more powerful and fundamental vision of gravity as the geography of spacetime does not break down in those circumstances, but general relativity predicts singularities of infinite density and spacetime curvature, and flounders at those infinities.

Black holes are a legacy of general relativity, but because they demonstrate so succinctly how gravity and light operate they have, in turn, provided a superb mental laboratory for exploring the most extreme implications of the theory, for finding out where this mathematical and conceptual framework leads us and where it fails us, for confronting the need to deepen the theory.

Again at the end of the twentieth century we grope for a new level of understanding, for a theory that doesn't break down where relativity theory does and that resolves the apparent inconsistency between it and another great twentieth-century physics theory – quantum mechanics. Once more, black holes provide a testing ground. When Hawking pondered the unwelcome possibility that black holes might emit radiation, he took a first step toward the unification of general relativity and quantum mechanics. We don't yet understand their 'fiery marriage', as Wheeler calls it, but we already have a name for it – 'quantum gravity'. Black holes have led us to this threshold. Here, they seem to hold keys even to knowledge about the origin and fate of the universe.

The theory of black holes was devised and developed for decades (centuries if we include their early proposal in the eighteenth century) by thought alone, with no astronomical observations to support it or serve as a guide. Now our technology allows us to look further into the cosmos with greater clarity than ever before, and we have found that our brain-child is not merely a sophisticated figment of imagination. Human minds and the mind of nature or God turn out to have been strangely and marvellously in synchrony. There *are* real black holes out there. On the astrophysical scale gravity actually does add up so as to overwhelm all the other forces, unleashing a monster mighty enough to extinguish the incredible detail of a star or group of stars – everything except mass, angular momentum, and charge – and to hold light itself prisoner.

How clever we've been, foreseeing the existence of black holes and figuring out so much about them with barely a hint from observational astronomy! However, now that we are able to study real black holes and the way they affect spacetime around them, we're also discovering, both to our chagrin and delight, that though the laws of gravity are simple, their elegant framework underlies a universe that is still full of astounding surprises. Our universe is dynamic, diverse, mysterious, and not nearly so predictable as we supposed. Even with our genius for theoretical extrapolation and the talent of a few of us for making brilliant intuitive leaps, it seems we humans haven't entirely second-guessed nature after all. We still must stand in awe.

Further reading

The following books provide good background reading and are all at about the same level of difficulty as *Prisons of Light*.

Chaisson, Eric. *Relatively Speaking: Relativity, Black Holes, and the Fate of the Universe*. London, New York: W. W. Norton & Co., 1988.

Relativity and its predictions explained in simple, clear language and analogies; a good portrait of Einstein himself.

Cornell, James, ed. *Bubbles, Voids, and Bumps in Time: The New Cosmology*. Cambridge: Cambridge University Press, 1989.

Vera C. Rubin's chapter is a particularly valuable introduction to the dark matter question.

Davies, Paul C. W. *The Search for Gravity Waves*. Cambridge: Cambridge University Press, 1980.

Especially strong on the background science necessary to understand this subject at the non-technical level.

Dodd, James E. *The Ideas of Particle Physics*. Cambridge: Cambridge University Press, 1988.

A serious, in-depth presentation, intended for use in school and university courses, but written in non-technical language, starting from the basics and assuming no prior knowledge of higher mathematics or physics.

Ferguson, Kitty. *Stephen Hawking: Quest for a Theory of Everything*. London: Bantam Press; 1992.

My book about Hawking's life and work, with simpler explanations of the science he wrote about in *A Brief History of Time*.

Feynmann, Richard P. *QED: The Strange Theory of Light and Matter.* Princeton: Princeton University Press, 1985.

Quantum electrodynamics – for everyone! There are books that merely describe scientific concepts to the lay reader, and there are books that allow lay persons to follow a scientist's thinking deeper into these concepts. Of the latter, this slim volume, taken from a series of public lectures,
is one of the easiest to follow, most worth the effort – and great fun.

Hawking, Stephen W. *A Brief History of Time: From the Big Bang to Black Holes.* London: Bantam Press, 1988.

Full of Hawking's humour and passion for his subject; and, in spite of the complaints levelled against it, not really all *that* difficult!

Hawking, Stephen W. *Black Holes and Baby Universes, and Other Essays.* London: Bantam Press, 1993.

A collection of lectures and articles Hawking has written for lay audiences, on various subjects, personal and scientific, over the period 1976 to 1992. Includes his excellent *Scientific American* article, 'The Quantum Mechanics of Black Holes', listed separately below.

Hawking, Stephen W. 'The Quantum Mechanics of Black Holes'. In *Scientific American*, Vol. 236 (No. 1), January 1977, pp. 34–40 and ff.

More than just the quantum mechanics of them, a concise, mostly non-technical review of theory having to do with black holes.

Longair, Malcolm. 'Our Evolving Universe'. Cambridge: Cambridge University Press, 1989.

A tour-de-force overview of what astronomers have discovered about the universe on the large scale since they have been able to study it in other wavebands besides the optical waveband; and an invaluable preparation for understanding what has been discovered even more recently.

Matzner, Richard, Tsvi Piran, and Tony Rothman. 'Demythologizing the Black Hole'. Also Matzner, Richard, Tony Rothman, and William Unruh. 'Grand Illusions: Further Conversations on the Edge of Spacetime'. Both in Tony Rothman, *Frontiers of Modern Physics*. New York: Dover Publications, Inc., 1985.

Entertainingly written, authorative and down to earth; featuring the computer simulation we saw on pages 85–7.

Thorne, Kip S. 'Gravitational Collapse'. In *Scientific American*, Vol. 217 (No. 5), November 1967, pp. 88–102 and ff.

The article that describes the collapse as we discussed it on pages 76–8.

Thorne, Kip. S. *Black Holes and Time Warps: Einstein's Outrageous Legacy*. London, New York: W. W. Norton & Co., 1994.

A top-notch, extremely thorough popular science book; showing science as a human process; particularly strong on the complementary roles theory and observation play in the evolution of scientific knowledge. Full of examples to support John Wheeler's comment that we are never more thoroughly convinced of something than when we have first fought it tooth and nail.

Wheeler, John Archibald. *Journey into Gravity and Spacetime*. New York: W. H. Freeman and Company, 1990.

From the mentor for generations of physicists and creator of the Princeton course, 'Physics for Poets'. No one explains more clearly or with more vivid analogies than Wheeler, or keeps the needs of his readers so constantly in mind.

Index